THE MANAGEMENT OF HAZARDOUS SUBSTANCES IN THE ENVIRONMENT

Selected papers presented at Envirotech Vienna 1989, First International ISEP Congress, Vienna, Austria, on 20–22 February 1989

Honorary committee

Federal Chancellor, Franz Vranitzky
Vice Chancellor and Minister for Foreign Affairs, Alois Mock
Federal Minister for Environment, Youth and Family, Marilies Flemming
Federal Minister for Science and Research, Hans Tuppy
Mayor of Vienna, Helmut Zilk
Vice Mayor of Vienna, Hans Mayr
Vice Governor of Lower Austria, Erwin Pröll
Municipal Counsellor for Environment, Leisure and Sports, Michael Häupl
Former Minister, Franz Kreuzer

Scientific committee

Jumpei Ando (Chuo-University/Tokyo, Chairman)
Winfried Lang (Univ. of Vienna/Co-chairman)
Frank A. Battig (Austrian Chemical Society, Vienna)
Werner Biffl (Univ. f. Bodenkultur, Vienna)
Winfried E. H. Blum (Univ. f. Bodenkultur, Vienna)
Pierre-Marie Dupuy (Univ. de Paris)
Kurt Fedra (IIASA, Laxenburg)
Volker Franzius (Federal Environmental Agency, Berlin)
Josef Frick (Austrian Association of Chemical Industry, Vienna)
Albert Hackl (TU Vienna)
J. F. K. Huber (Univ. of Vienna)
N. Kochupillai (All India Inst. of Medical Sciences, New Delhi)
S. H. Mikhaylov (Sofia)
James Mwalusanya (High Court of Tanzania)
Howard T. Odum (Univ. of Florida)
Rupert Riedl (Univ. of Vienna)
Heinz Schreiber (Federal Ministry of Environment, Youth and Family, Vienna)
Uri Shamir (Technion–Israel Institute of Technology, Haifa)
Jinzaburo Takagi (Citizens' Nuclear Information Center, Tokyo)
Frederic Vester (Study Group for Biology and Environment, Munich)
Siegfried Waldhör (Henkel Austria GmbH, Vienna)
Kurt Wedl (Academy of Environment and Energy, Laxenburg)
Wolfgang Wieser (Univ. of Innsbruck)
Klaus Wolf (Municipal Environmental Agency, Hamburg)
Konrad L. Zirm (Federal Environmental Agency, Vienna)

Organizing committee

Rudolf Kadanka, Raoul Kneucker, Johannes Mayer, Trude Welan, Konrad L. Zirm (International Society for Environmental Protection/ISEP)
Gerd Hoffmann (Wiener Messe AG)
Annette Jirasek (MONDIAL CONGRESS)
Christian Mutschlechner (Vienna Convention Bureau)
Walter Straub (Vienna Airport Authority)

THE MANAGEMENT OF HAZARDOUS SUBSTANCES IN THE ENVIRONMENT

Edited by

K. L. Zirm *and* **J. Mayer**

Published on behalf of the

INTERNATIONAL SOCIETY FOR ENVIRONMENTAL PROTECTION

by

ELSEVIER APPLIED SCIENCE

LONDON and NEW YORK

ELSEVIER SCIENCE PUBLISHERS LTD
Crown House, Linton Road, Barking, Essex IG11 8JU, England

Sole Distributor in the USA and Canada
ELSEVIER SCIENCE PUBLISHING CO., INC.
655 Avenue of the Americas, New York, NY 10010, USA

WITH 8 TABLES AND 29 ILLUSTRATIONS

British Library Cataloguing in Publication Data
The management of hazardous substances in the environment.
1. Environment. Pollution
I. Zirm, K. L. II. Mayer, J. III. International Society for Environmental Protection
363.73

ISBN 1-85166-459-9

Library of Congress Cataloging-in-Publication Data
The management of hazardous substances in the environment / edited by K. L. Zirm, J. Mayer.
p. cm.
'Selected papers presented at Envirotech Vienna 1989, First International ISEP Congress, Vienna, Austria, on 20–22 February 1989'—
Includes bibliographical references.
ISBN 1-85166-459-9
1. Hazardous wastes—Management—Congresses. I. Zirm, K. L. II. Mayer, J. III. International Society for Environmental Protection. IV. Envirotech Vienna 1989 (1989) V. International ISEP Congress (1st : 1989 : Vienna, Austria)
TD1030.M36 1990
363.72'87—dc20 90-31083
CIP

Printed by The Universities Press (Belfast) Ltd.

Preface

There has been growing international concern in recent years about the urgent environmental problems associated with hazardous substances and their management. Discussion has focussed on three main areas:

(1) *Evaluation of hazardous substances* arising either as products or by-products of modern industrialised society.
(2) *Current developments in environmental technology* designed to reduce the level of hazardous substances released into the environment.
(3) *Effective measures of waste management* designed to treat hazardous substances present in the environment and to detect and clean up contaminated sites.

This volume contains selected keynote papers covering these major themes, which were presented at Envirotech Vienna 1989, organised by the International Society for Environmental Protection. Additional volumes containing session papers are available separately (see Appendix II).

Further discussion of these issues is planned at Envirotech Vienna 1990 which will be held at the Austria Center, Vienna, on 23–27 October 1990. The theme of the conference will be 'Current Problems in Hazardous Waste Management and Contaminated Sites'. Abstracts of proposed papers and requests for further information should be sent to: International Society for Environmental Protection, A-1090 Vienna, Spittelauer Lände 3, Austria. Telephone: 43 222 34 45 11 510. Telefax: 43 222 34 45 11 299.

Contents

List of Contributors

FERDINANDO ALBANESE

Director of Environment and Local Authorities of the Council of Europe, Boîte Postale 431 R6, F-67006 Strasbourg cedex, France

JUMPEI ANDO

Faculty of Science and Engineering, Chuo University, Kasuga, Bunkyo-ku, Tokyo 112, Japan

HANS ULRICH BAMBAUER

Institut für Mineralogie, Universität, Corrensstrasse 24, D-4400 Münster, FRG

FINN BRO-RASMUSSEN

Laboratory of Environmental Science and Ecology, The Technical University of Denmark, DK-2800 Lyngby, Denmark

HELLMUT FLECKSEDER

TU Wien, Institut für Wassergüte und Landschaftswasserbau, Karlsplatz 13, A-1040 Vienna, Austria

N. GALIL

Environmental and Water Resources Engineering, Technion—Israel Institute of Technology, Faculty of Civil Engineering, Technion City, IL-32000 Haifa, Israel

GEORG GEBHARD

Rheinisch-Westfälisches Elektrizitätswerk, Kruppstraße 5, D-4300 Essen, FRG

G. HINTEREGGER

Executive Secretary, United Nations Economic Commission for Europe, Palais des Nations, CH-1211 Geneva, Switzerland

THOMAS HOLZAPFEL

Rheinisch-Westfälisches Elektrizitätswerk, Kruppstraße 5, D-4300 Essen, FRG

L. JOURDAN

CEFIC, 250 Avenue Louise, 1050 Bruxelles, Belgium

MARGARET M. KELLY

Director, Technology Staff, Office of Program Management and Technology, Office of Solid Waste and Emergency Response, The United States Environmental Protection Agency, 401 M Street, SW, Washington, DC 20460, USA

WALTER KLÖPFFER

Battelle-Institut e.V., Abteilung Chemische Analytik, Am Römerhof 35, D-6000 Frankfurt am Main 90, FRG

CHRISTIAN KRAUSE

Institut für Mineralogie, Universität, Corrensstrasse 24, D-4400 Münster, FRG

WALTER LUKAS

Institut für Baustofflehre und Materialprüfung, Universität, Technikerstraße 13, A-6020 Innsbruck, FRG

D. C. McBAIN

Scientific Affairs Officer, International Register of Potentially Toxic Chemicals, United Nations Environment Programme, Palais des Nations, 1211 Geneva 10, Switzerland

RONALD A. N. MCCLEAN
Arthur D. Little Ltd, Science Park, Cambridge CB4 4DH, UK

M. REBHUN
Environmental and Water Resources Engineering, Technion—Israel Institute of Technology, Faculty of Civil Engineering, Technion City, IL-32000 Haifa, Israel

MARIE-CLAUDE SAVOIE
Arthur D. Little Inc. (Canada), 67 Yonge Street, Toronto, Ontario, Canada

MASARU TANAKA
The Institute of Public Health, Ministry of Health and Welfare, 6-1 Shiroganedai 4-chome, Minato-ku, Tokyo 108, Japan

GUDRUN WILLNER
Institut für Mineralogie, Universität, Corrensstrasse 24, D-4400 Münster, FRG

HARVEY YAKOWITZ
Consultant to the Environment Directorate, Organisation for Economic Cooperation and Development (OECD), 2 Rue André Pascal, F-75775 Paris cedex 16, France

1

Environmental Policy: Strategies and International Cooperation

G. HINTEREGGER

Executive Secretary, United Nations Economic Commission for Europe, Palais des Nations, CH-1211 Geneva, Switzerland

This book is published at a critical moment. There is a growing awareness in public opinion as well as among policy-makers that the degradation of the environment with its negative economic and social consequences is assuming dramatic proportions, and that urgent measures are needed at all levels, national, regional and global, in order to avoid an environmental catastrophe.

The Earth's atmosphere is being changed at an unprecedented rate by pollutants resulting from a multitude of human activities. Climatic warming, depletion of the ozone layer, effects on vegetation, soil, water and other environmental resources as a result of long-range transport of pollutants including acidifying substances, threaten our very future. Much of this damage is avoidable, in particular the inefficient and wasteful use of natural resources and energy. Pollution of air, soil and water and the danger connected with hazardous wastes are only some of the multifaceted problems which confront governments today and demand concerted action for their resolution.

There has been an impressive response by the international community to these challenges both in terms of strategic consideration as well as in concrete measures of concerted action; but of course a lot remains to be done in the future. At the global level it is the United Nations Environment Programme (UNEP) which, in cooperation with other United Nations agencies, including the United Nations Economic Commission for Europe (ECE), is mandated to find solutions for the most pressing issues in that field. A global Environmental Strategy elaborated under the auspices of UNEP was adopted by the UN

General Assembly in 1987. The Report of the World Commission on Environment and Development (WCED), also referred to as the report of the Brundtland Commission, introduced the concept of 'sustainable development' which is now widely accepted as a guiding principle for the integration of environmental considerations to all economic activities and social policies. Most recently climate change has been declared a priority issue for international cooperation by the UN General Assembly. As a direct follow-up to the 1985 Vienna Convention for the Protection of the Ozone Layer, established under the auspices of UNEP, a Protocol was adopted and signed in Montreal in 1987 that calls for a freeze of controlled chlorofluorocarbons (CFCs) as of 1 January 1989 at their 1986 level and sets 50% reduction in production and consumption of CFCs by mid-1998. At present, preparations for a global convention on the control of transboundary movements of hazardous wastes, in which ECE experts participate actively, have reached a critical point which hopefully can be overcome so that the convention can be adopted in the near future.

This chapter focuses specifically on regional strategies and cooperation regarding environmental matters which are pursued actively within the Economic Commission for Europe, one of the five regional commissions of the United Nations. Membership consists of 34 countries comprising all the market-economy countries of both Europe and North America as well as the European centrally planned-economy countries. Indeed, for over 40 years the Commission has served as a unique forum for East–West cooperation, among others in the field of environment.

While the ECE region is marked by considerable diversity with regard to levels of economic development and natural-resource endowment—as well as political systems—it collectively shares a number of important socio-economic characteristics and common environmental assets. Together, the ECE region represents more than two-thirds of total world Gross Domestic Product (GDP) and two-thirds of world trade. It includes countries with the highest standard of living in the world as measured by per capita GDP. A great share of modern technology is being elaborated, developed, manufactured, and marketed by and in the countries of the ECE region. As for the other side of the coin, the ECE region accounts for more than three-quarters of world pollution and is the heaviest consumer of natural resources and energy. Seventy per cent of world emissions of sulphur dioxide, 60% of nitrogen oxides and more than 80% of the total chlorofluorocarbons are released in countries of the region. This places a heavy responsibility on ECE countries for environmental protection, both in a regional context and with regard to their contribution to the solution of global problems like climate change.

The UN ECE is well-equipped to deal with environmental issues both separately and in a broader context. Owing to its intergovernmental, cross-sectoral and interdisciplinary nature—the Commission is engaged in ten major programmes dealing with economic issues: economic projections; trade; science and technology; industry; agriculture; timber, environment; transport; energy and human settlements—member governments are able to develop a particularly comprehensive programme dealing with environmental concerns under ECE auspices. The specific bodies carrying out this work are: the Senior Advisors to ECE Governments on Environmental and Water Problems; and the Executive Body for the implementation of the Convention on Long-range Transboundary Air Pollution. However, most of the other Principal Subsidiary Bodies of the Commission are also involved because environmental considerations are linked to virtually all areas of economic activity and thus touch nearly all activities of the Commission.

In our region the response of governments to the challenges of environmental degradation has been particularly energetic and productive, both in terms of strategies and policy recommendations and in terms of multilateral regulatory instruments of a legally binding character. It is worth noting that East–West cooperation in the area of environment has quickly developed into one of our priority activities as environmental concerns are now shared by all member countries of the ECE. This common concern has made possible the formulation of a joint strategic concept.

The ECE adopted, at its 43rd session in 1988, the Regional Strategy for Environmental Protection and Rational Use of Natural Resources covering the period up to the year 2000 and beyond. The Strategy manifests the view of ECE governments that protection of the environment and rational management of natural resources are integral to economic and social development. It further reflects their perception that continuing and emerging environmental and resource management problems, as well as new scientific, technological and institutional opportunities, can benefit substantially from closer, better-focused collaborative action by ECE member countries over the near and longer terms. It aims at the eventual creation in the ECE region of a situation whereby economic activity and social development can proceed in an environment essentially free from: the degrading effects of pollution; the threat of hazards to human health posed by chemicals and wastes; loss of values and opportunities associated with a broad and stable natural-resource base; and disagreements over environmental issues of a transboundary nature. The Strategy supports the development of technologies and practices for a better environment. It also establishes a series of policy and programme responses to achieving these goals.

A key element of such policy is integration of environmental considerations into all economic activities. In view of the high degree of interdependency which exists in the field of environment—interdisciplinary, cross-sectoral, national, international—governments will continue to face integration problems when defining and implementing their policies and strategies. It is well-recognized that problems in one area are often linked to others and can only be resolved by changes in still other areas of concern: a sort of chain reaction within a multidimensional system. Acid deposition, and its association with environmental conditions in air, water and land, is a prime example of problems that are interconnected. That is why a horizontally integrated, multimedia policy approach must be taken so as to avoid transferring pollution from one environmental medium to another.

The 'economy–ecology' integration aspect of environmental policies and strategies has been highlighted in the Report of the World Commission on Environment and Development (WCED). Governments are beginning to recognize that they have to secure a dynamic equilibrium in the relationship between environmental quality, rational use of natural resources and sustainable economic development. The concept of sustainable development defined by WCED as 'development that meets the needs of the present without compromising the ability of future generations to meet their own needs', could be interpreted as meaning that if socio-economic development is to be pursued, its continuation and expansion depend critically on mankind's ability to maintain the resource base which makes this development possible; otherwise degradation will impinge on the quality of life in the long run. This concept will undoubtedly become a cornerstone of global and regional environment policy.

In addition to the Regional Strategy a considerable number of other documents of a strategy nature or containing policy recommendations have been elaborated and adopted in the framework of the UN Economic Commission for Europe.

In 1988 the ECE adopted the Declaration on the Conservation of Flora, Fauna and their Habitats. This Declaration embodies a conceptual approach to international cooperation on nature protection. It makes specific provisions for government action needed to promote sustainable use of biological resources in the ECE region in the interest of present and future generations.

Governments working within the ECE have also developed and implemented policies and strategies for integrated water management. Declarations and recommendations elaborated to this end include the ECE Declaration of Policy on the Rational Use of Water, adopted by the Commission in 1984. The Declaration calls for the rapid development and sound application of appropriate technology and its efficient operation to ensure wasteless use of precious natural resources. The Commission also adopted in 1988 recommendations to ECE governments on wastewater treatment.

An ECE Charter on groundwater management is now being finalized. It draws the attention of governments and alerts public opinion to the need for sustainable use of groundwater and the preservation of the vital quality of groundwater resources, often non-renewable and which by their very nature are hidden from view and risk mismanagement.

Problems associated with equitable use of shared water resources, in particular, prevention and control of pollution in transboundary waters, have increasingly assumed special importance. ECE governments have recognized the urgency of transboundary water issues and the need for jointly agreed principles for enhanced cooperation among riparian states. The ECE adopted the ECE Declaration of Policy on Prevention and Control of Water Pollution Including Transboundary Pollution, a decision on International Cooperation on Shared Water Resources and, more recently, Principles Regarding Cooperation in the Field of Transboundary Waters. The Principles address problems of prevention and control of transboundary water pollution as well as flood management. They call for special attention to be given to hazardous substances, especially those which are toxic, persistent, bioaccumulative, and whose introduction into transboundary waters should be prohibited if not prevented through the use of the 'best available' technology.

In response to a number of recent emergencies, a Code of Conduct on accidental pollution of transboundary inland waters is now being elaborated. The Code will contain jointly agreed procedures whereby competent authorities in individual member countries can take coordinated steps to prevent and control transboundary accidental pollution. It will help improve preparedness-planning to cope with emergencies; arrange for mutual information exchange and assistance; and apply measures to contain, restore and compensate for damage resulting from accidental pollution in transboundary waters.

The UN Economic Commission for Europe also has an impressive record of achievements regarding the elaboration of legally binding multilateral instruments and other regulatory measures concerning the protection of the environment.

In the mid-1970s, the work of the Commission in the field of the environment gained substantial momentum as a result of environmental provisions set down in the Final Act of the Conference on Security and Cooperation in Europe (CSCE) signed in Helsinki in 1975; and complemented in 1983 by the Concluding Document of the Madrid Meeting of the participating states of CSCE. Not only were traditional areas of cooperation strengthened but new areas of cooperation within the ECE were identified. New environmental initiatives were undertaken with a high degree of success under the broader mandate. The Final Act of CSCE gave impetus to the High-level Meeting of 1979 within the Framework of the ECE on the Protection of the Environment. Representation at a very high level, including many Ministers, enabled the meeting to adopt the Convention on Long-Range Transboundary Air Pollution, the first multilateral legal instrument in this field, and the Declaration on Low- and Non-Waste Technology and Reutilization and Recycling of Wastes.

At present, 31 states and the European Economic Community have ratified the Convention on Long-Range Transboundary Air Pollution and actively participate in its implementation. The Convention embodies a far-reaching mechanism for dealing with specific environmental concerns. Supplementary protocols to the Convention cover a range of specific subjects: in 1984, the Protocol on the long-term financing of the Cooperative Programme for Monitoring and Evaluation of the Long-Range Transmission of Air Pollutants in Europe (EMEP) was adopted and has since been ratified by 31 parties. The following year in Helsinki a Protocol on sulphur dioxide was adopted, and entered into force for 18 countries in 1987. The parties undertake to reduce their national annual sulphur emissions or their transboundary fluxes by at least 30% by the year 1993 on the basis of the 1980 level of emissions.

More recently, on 1 November 1988 in Sofia at the meeting of the

Executive Body of the Convention in which numerous environment ministers participated, 25 parties to the convention signed the Protocol concerning the control of emissions of nitrogen oxides or their transboundary fluxes, with a package of internationally agreed abatement measures starting with a 'freeze' of emissions from 1994. In addition, 12 countries signed a Declaration committing them to a 30% reduction of their national emissions of nitrogen oxides by 1998. This meeting also mapped out future activities under the Convention regarding emission control for hydrocarbons (volatile organic compounds); abatement strategies taking into account the concept of critical loads; and exchange of technology for air pollution control.

The real impact of the Convention in terms of overall pollution trends is gratifying. A recent Review of National Strategies and Policies for Air Pollution Abatement showed that at least 23 member states had achieved a net reduction of sulphur emissions as compared to 1980. Ten countries had actually reduced their emissions by 30% or more, even before the entry into force of the Helsinki Protocol in September 1987. Overall emissions of sulphur dioxides in Europe have already dropped by 15% between 1980 and 1986. Eleven countries have announced long-term national reductions of about 50% or more, four of them even more than 65%. While governments can thus look to their work through the Convention with some degree of pride, this effort must, of course, continue and expand.

Increased attention is also being paid to environmental impact assessment as an important planning tool for preventing and controlling adverse environmental effects. A framework agreement is being developed to regulate the application of environmental impact assessment in a transboundary context. This ECE agreement will constitute an internationally, legally binding instrument stipulating commonly agreed provisions to carry out jointly or to coordinate national procedures to assess environmental impacts and to arrange for the application of environmental impact assessment at an early stage of planning for any activities likely to cause transboundary environmental impacts.

Another area where legal instruments relevant to environmental concerns have been or are being elaborated in the ECE is that of transport of dangerous goods where the ECE ensures international cooperation through the implementation of relevant European Agreements. Its provisions and recommendations cover all transport modes not only in our region but throughout the world. A draft Convention on civil liability and damage caused during carriage of dangerous goods has just been elaborated by the ECE Inland Transport Committee and is expected to be signed in the near future.

The second major achievement of the High-level Meeting in 1979 in addition to the Convention on Long-Range Transboundary Air Pollution was the Declaration on Low- and Non-Waste Technology and Re-utilization and Recycling of Wastes. The pioneering role of the ECE in this area deserves particular mention. Uncontrolled amounts of hazardous wastes, dumped into the environment or on ill-chosen landfill sites have caused serious problems. A solution to this difficult problem is being sought through policies and strategies to minimize the generation of hazardous waste. Waste reduction is promoted through industrial processes that yield either very little waste or preferably none at all. This approach is based on abatement of pollution at the source. The Declaration calls also for the use of energy- and resource-saving technologies as a means to reduce the amount of waste generated per unit of product.

While low- and non-waste technology will continue to be the cornerstone of policies and strategies to deal with the problem of hazardous wastes within the ECE, the shortfall between realization of that objective and the current situation must nevertheless be faced. Recovery of residual material, recycling and re-utilization of hazardous wastes; their effective and environmentally sound treatment to reduce waste volumes, detoxification and/or destruction of wastes, safe storage or temporary disposal for re-use—these are all among the options which may be applied in place of land disposal of hazardous wastes with inherent risks to man and the environment. Hazardous wastes must be handled in a coordinated, effective manner or serious problems and dangers arise.

The Compendium on Low- and Non-Waste Technology, originally compiled by the ECE contains more than 160 examples of environmentally sound technologies applied on an industrial scale. The Compendium has now been entrusted to the Geneva-based International Environmental Bureau of the International Chamber of Commerce for further action, with the intention of profiting from its comparative advantages of close contacts to industry and business at a global level.

The Commission is promoting international cooperation regarding environmentally sound technology in various ways. By its decision on cooperation in the field of environmental protection and water resources, the ECE invited member states to consider improving the use of effective and environmentally sound technologies, taking into account the necessity of cooperating effectively in preventing and combating pollution so as to preserve and improve the environment. To this end the Commission has asked member states to promote the commercial exchange of available technology as well as direct industrial contacts in the field of environmental protection. This is and will remain one of the pillars of modern environmental strategy.

The ECE Convention on Long-Range Transboundary Air Pollution deals with some of the questions of technology. In the design of air pollution management systems it calls for control measures to be developed using the best available technologies which are economically feasible. Several of our seminars on air pollution control technology have contributed to the exchange of information and know-how in this field. The Executive Body for the Convention has also recently established a task force on the exchange of technology, in order to assist parties to the Convention to meet their obligations under the Convention and its protocols.

In the context of a joint project with the United Nations Development Programme, the ECE has already initiated specific programmes for technical information exchange and industrial cooperation in the fields of flue gas treatment, fuel treatment, low-emission combustion processes and by-products utilization.

Minimization of wastes, in particular hazardous waste, through the application of low- and non-waste technology, needs to be incorporated in integrated waste management taking full account of the whole life-cycle of wastes—generation, collection, storage, treatment, reutilization and/or final disposal. This comprehensive approach—the cradle-to-grave approach—would encompass environmental protection strategies with development of the economy in general and industry in particular. This represents another facet of integration that could result in the elaboration of a regional strategy for integrated waste management, with particular emphasis on hazardous wastes. A concerted systematic approach to waste management is absolutely necessary in order to resolve environmental problems associated with the generation of hazardous wastes in production and consumption processes and with the environmentally sound handling and disposal of hazardous wastes in Europe and North America.

We are all only too aware of the industrial accidents with transboundary environmental implications that have occurred recently.

Among other things they highlight the significance and wide range of adverse environmental effects. These emergencies have also revealed a certain degree of unpreparedness. Moreover, existing international legislation and the usual procedures appear to be inadequate to cope with such emergencies. The frequency and severity of such accidents is predicted to increase, along with expenditures for damage-mitigation, containment and rehabilitation. Risk of major accidents also increases as technology becomes more and more sophisticated, complex and less tolerant of human failure in operation. Here again there would seem to be no choice but to take adequate measures coordinated at an international level. International regulations drawn up so far concern industrial sectors or specific fields of international cooperation alone. There is now sufficient need for countries to define through specific agreements their relations and conduct in case of accidents with adverse transboundary environmental effects.

It is hoped that this comprehensive presentation of ECE activities in the field of the environment has shown clearly that there is indeed an impressive record of achievement as a result of regional cooperation. However, this must not be a reason for complacency as a lot remains to be done. What is encouraging is the clear manifestation of a strong political will for international cooperation in the field of environment that found its most recent expression in the Concluding Document of the Vienna Follow-up Meeting to the Conference on Security and Cooperation in Europe. This document not only contains concrete proposals for cooperation in the area of environment; it also stipulates the convening of a Meeting on the Protection of the Environment held in Sofia (Bulgaria) from 16 October to 3 November 1989. The UN Economic Commission for Europe participated in this Meeting and contributed a report containing recommendations for concrete action in each priority area: industrial accidents with transboundary impacts; potentially hazardous chemicals; and transboundary water pollution. A clear understanding was reached that action should be taken by the ECE as soon as possible, to elaborate a legal instrument for the prevention and control of the transboundary effects of industrial accidents, as well as the elaboration of a framework convention on the use and protection of transboundary water courses and international lakes.

In addition the Government of Norway, in close cooperation with the ECE, organized a regional conference on the environment and sustainable economic development held in Bergen (Norway) from 8 to

16 May 1990, with representation at the Ministerial level. This Conference reviewed progress made on implementation of selected aspects of the WCED report and identified initiatives for further measures to translate the concept of sustainable development into a plan of action within the regional context. One of the preparatory meetings to this Conference dealt with the 'Economics of Sustainability'—in other words, the planning and management of economic activity as a whole must take into account environmental considerations—was held in the United States in autumn 1989. This was also one of the four key areas which the Conference focused on, the others being 'sustainable energy use', 'sustainable industrial activity' and 'awareness raising and public participation'.

2

The Contribution of the Council of Europe Towards the Elaboration of a Policy and of Instruments for the Protection of the Environment

FERDINANDO ALBANESE

Director of Environment and Local Authorities of the Council of Europe, Boîte Postale 431 R6, F-67006 Strasbourg cedex, France

INTRODUCTION

For 25 years the Council of Europe has been striving to stimulate the policies of its member states on matters concerning the environment. Whilst the Council of Europe has concenrated its own action on nature conservation, problems concerning the protection and management of natural resources have not been neglected.

Problems posed by hazardous chemicals and toxic waste have been dealt with in an indirect rather than direct manner in so far as pollution caused by chemicals is taken into consideration when dealing with more general questions. This is the case, for example, with the Draft European Convention for the Protection of International Watercourses Against Pollution which, unfortunately, has not been completed, as well as with the study currently being conducted by our organisation on soil protection.

Thus, I should like to give a brief account of the action carried out by the Council of Europe towards the elaboration of a national and international policy for the protection of the environment. The objective in this regard has been to manage the natural environment in the most appropriate way possible, while taking into account human requirements and necessities on both a social and an economic level.

POLICIES

In 1979, in Bern, the Convention on the Conservation of European Wildlife and Natural Habitats—otherwise known as the 'Bern

Convention'—was opened for signature. Its aim is to protect threatened species of flora and fauna and their habitats, to protect endangered natural habitats of migratory species, and to safeguard migratory species themselves. The protection of migratory species adds a dimension of north–south interdependency to the Convention. Indeed, on top of the 17 member states, together with Finland and the European Community, Senegal is also bound by the Convention, while Hungary, Tunisia and Morocco have been invited to accede.

As regards the means set out by the Convention for the protection of flora and fauna, the following should be mentioned:

(a) for strictly protected flora species: prohibition of deliberate picking, gathering, cutting or up-rooting of plants, and the protection of their habitats;
(b) for strictly protected fauna species: prohibition of deliberate capture, destruction of breeding or resting sites, deliberate disturbance of fauna, deliberate destruction or taking of eggs, destruction of or trade in these species;
(c) for species of protected fauna: ensuring that endangered species be kept out of danger by introducing closed seasons for hunting, prohibiting temporary or local exploitation, regulating the sale, keeping for sale or transporting for sale, prohibiting certain hunting methods.

A list of these protected species of flora and flauna is to be found in the Appendices to the Convention. It should be remembered that invertebrate species and freshwater fish have recently been added to the Appendices which now cover the whole of the animal kingdom.

Ever since the negotiating stage of the Convention, the Council of Europe has been aware of the need to integrate nature conservation into national social and economic policies. Under Article 3(2) of the Convention, each state undertakes, in its planning and development policies and in its measures to fight against pollution, to give due regard to the conservation of wild flora and fauna.

Thus, the Bern Convention constitutes the legal framework for nature conservation policy while at the same time providing a management and protection instrument, in so far as a Standing Committee, made up of representatives of the contracting parties, monitors the application of the Convention by drawing up appropriate recommendations.

The organisation is striving to broaden and improve protection policies by elaborating a Conservation Strategy for Europe, as part of a worldwide strategy, thereby taking up a suggestion made at the 5th European Ministerial Conference on the Environment (Lisbon, June 1987). It is planned that the Strategy should become a source of inspiration for the actions of governments and European organisations.

The document, currently available in draft form, contains, in particular, chapters on air, surface and ground water, the sea, soil, hazardous chemical and nuclear waste.

With regard to the latter of these, the following action has been considered necessary:

(a) Seek to reduce levels of waste generation by careful husbandry and encourage re-use and recycling.
(b) Provide guidelines nationally and internationally to overcome the considerable technical problems in handling, transporting and storing hazardous wastes, especially from the generation of nuclear power, and to develop improved procedures and technologies for the environmentally sound management of hazardous wastes.
(c) Set common standards for maximum acceptable levels in human tissues and contamination sources, and for permitted uses and exposures.
(d) Develop early warning and emergency procedures, nationally and internationally, to mitigate the effects of accidental discharges.
(e) Support efforts of international organisations with respect to the exchange of information about toxic substances.
(f) Ensure that new chemical products are assessed not just in relation to hazards to people but also with regard to their effects on wildlife before allowing such chemicals to be marketed and used.
(g) Ensure that transnational corporations comply with legislation regarding hazardous and toxic substances in both their home and host countries.

The aim of the Conservation Strategy for Europe is to formulate a comprehensive and coherent approach to all environmental problems. In this way, the Strategy could continue action already started by the Council of Europe to draw up specific protection policies, which led to the adoption of a series of recommendations and charters on natural resource management, of which the following should be mentioned: the Declaration of Principles on Air Pollution Control (Resolution (68) 4), the European Water Charter (Resolution (67) 10), the Protection of Lake Shores and River Banks (Resolution (77) 8), the Protection of Coastal Areas (Resolution (73) 29), the European Soil Charter (Resolution (72) 19), the Ecological Charter for Mountain Regions in Europe (Resolution (76) 34). Other studies and texts are available which deal with matters such as the protection of habitats and threatened species of flora and fauna, as well as dealing with the protection of landscapes. A feasibility study is currently being conducted on the possibility of devising legal instruments for soil protection, while also looking at the need to protect groundwater.

This feasibility study will be completed for use at the 6th European Ministerial Conference on the Environment, to be held in Belgium in 1990, which will be putting forward proposals on action to be taken in this field.

The study on soil protection, in the current text drawn up by a consultant expert, deals first with the concept and function of soil, and then examines the causes of soil degradation as well as national and international measures taken for its protection. Proposals for action by the Council of Europe have been drawn up. One chapter, in particular, studies problems caused by chemicals.

Water protection is the subject of a draft Convention which, due to oppositions expressed within the Committee of Ministers, has not yet been opened for signature. Nevertheless, I think it would be useful to give a brief summary of its content in order to illustrate an example of an efficient method that could be used for regulating pollution problems in international watercourses.

The Draft European Convention for the Protection of International Watercourses against Pollution seeks to establish both: (a) water quality objectives and (b) a method for attaining these objectives.

Objectives

Under the Convention, each interested party shall strive to protect all surface water within its territory, using any measures that would: (a) reduce existing levels of pollution and (b) prevent new forms of pollution. Where international watercourses are concerned, the obligations to which interested parties are committed include the gradual reduction in the general level of existing pollution and prevention of new forms of pollution. Interested parties shall go about their obligations in two fundamental ways: (i) maintaining or bringing the quality of international watercourses to levels not less than those values which each party shall set using the Appendix as their basis, and according to the different possible uses of the water; and (ii) seeking authorisation before discharging any of the substances listed in Appendix II into the waters of international hydrographic basins.

Methods

To achieve these objectives the Convention looks to cooperation between interested parties. This cooperation should take place in two ways: (a) setting up an information and warning system in case of a sudden increase of pollution in international watercourses and (b) forming international commissions, for the relevant hydrographic basins, by way of cooperation agreements, within which the parties undertake to enter into negotiation with each other.

The essential function of these international commissions is to propose to the interested parties the assignment of the international watercourse under its authority, or one or more of its sections, to one or more possible uses, namely:

(a) production of drinking water for human consumption;
(b) consumption by domestic and wild animals;
(c) the conservation and enhancement of wildlife, both flora and fauna, and the preservation of the self-purifying capacity of water;
(d) fishing;
(e) recreational purposes, regard being had to health and aesthetic requirements;
(f) the application of freshwater directly or indirectly to land for agricultural purposes;
(g) the production of water for industrial purposes;
(h) the need to preserve an acceptable quality of sea water; and
(i) the conservation of the aquatic environment in the regions concerned.

On the basis of these uses, the international commissions devise water quality objectives and the detailed rules for applying them which are then proposed to interested parties. To set these objectives, the commissions work according to those quality objectives defined in Appendix III of the Convention.

In the absence of the water quality objectives contained in Appendix III, the minimum values laid down in Appendix I of this Convention may serve as a basis for these consultations. It is important to note that these minimum values are to be applied:

(a) in the case of the freshwater values, at the freshwater limit and at each point upstream from this limit where the watercourse is crossed by a frontier between states;
(b) in the case of brackish water values, at the baseline of the territorial sea and at the points where the estuary is crossed by a frontier between states.

A system designed to settle disputes is included in case either: (i) the international commission fails to agree on the adoption of a proposal, or (ii) one of the interested parties to the agreement does not approve a proposal submitted to it by the international commission of which it is a member.

INSTRUMENTS

As regards the instruments of management and protection, several have already been developed or are in the process of being developed by the Council of Europe.

The establishment of the network of biogenetic reserves, on the basis of Resolution (76) 17 adopted by the Committee of Ministers, has the purpose of setting up nature reserves throughout the European Continent, which would contain representative specimens of different types of fauna, flora and biotopes of European origin, thereby securing their breeding. The network presently comprises approximately 150 reserves of this kind.

The European Diploma, created in accordance with Resolution (73) 4 adopted by the Committee of Ministers, is awarded to territories, sites or natural monuments of international value and special European interest from a natural heritage point of view, their being of scientific, cultural, aesthetic or recreational quality and provided that they already enjoy an adequate system of protection. Tenable for a renewable period of 5 years, this award gives the Council of Europe a certain degree of control in that it reserves the right to withdraw the Diploma if deterioration in the quality of the site is allowed to occur. Thirty Diplomas have, to date, been awarded.

Education and information: measures of protection and management for natural heritage would not work efficiently without proper support from interested groups. Therefore, the Council of Europe has been leading a public awareness and education campaign aimed at specific groups such as farmers, hunters, local residents and decision-makers from the protected area as well as the media. As part of this campaign, a Documentation and Information Centre on Nature and the Environment has been established within the Council of Europe. Expert groups as well as the general public are kept in touch with developments in the Council of Europe and its member states by way of a series of publications, information leaflets and periodicals *Newsletter—Nature—Environment* and *NATUROPA*, which appear in several languages, are two important examples of periodicals worth mentioning.

Another instrument capable of enhancing the efficiency of environmental protection while at the same time enabling it to be carried out in a democratic fashion is that of citizen participation in the decision-making process on matters concerning the environment.

On 17 September 1987 the Committee of Ministers adopted Recommendation R (87) 16 on administrative procedures affecting a large number of persons. Although this recommendation covers more than simply the environment (it covers all non-normative administrative acts), it is generally felt that the environment is the area most likely to benefit from its application.

The Recommendation is based on the observation that a growing number of interventions by administrative authorities are of such a far-reaching and complex nature that they affect in varying ways large numbers of people. Even parts of neighbouring states can be affected. The fundamental principle laid down by the Recommendation is that all administrative acts that affect large numbers of persons should be taken, on completion of a participation procedure. Consequently, when a competent authority proposes to take such an administrative act, the persons concerned should be informed in such a manner as may be appropriate and be provided with such factors as will enable them to judge its possible effects on their rights, liberties and interests.

The persons whose rights, liberties and interests are liable to be affected by the administrative act are grouped into three categories:

(a) the persons to whom the administrative act is addressed;
(b) persons whose rights, liberties or interests are liable to be

affected by the administrative act even though it is not specifically addressed to them; and

(c) a large number of persons who, under national law, have the right to claim a specific collective interest that is likely to be affected by the administrative act.

Taking these categories into account, the Recommendation proposes a number of principles to governments concerning:

(a) the organisation of consultation procedures, also involving, if necessary, the persons living in neighbouring states;
(b) the provision of access, first, to information to facilitate the participation procedure and, second, to the main conclusions emerging from the procedure; and
(c) information on normal remedies against the administrative act and the time-limit within which they must be utilised.

Finally, another instrument for protection, currently being examined at the Council of Europe, concerns a text on civil liability for damage caused to the environment. Very delicate problems remain to be solved in this area: for example, establishing a legal definition for environment, determining what is meant by notions such as damage, liable persons, persons entitled to damages, causal links, and compensation.

Despite these difficulties, the Council of Europe hopes to reach a satisfactory solution. Indeed, we are convinced that the insertion of stricter liability for damage caused to the environment into our laws will have a preventative effect and, thus, as in other branches of law (such as product liability) lead to greater protection of the interests in question.

3

Chemical Assessment, Information Exchange and the International Register of Potentially Toxic Chemicals (IRPTC)

D. C. McBain

Scientific Affairs Officer, International Register of Potentially Toxic Chemicals, United Nations Environment Programme, Palais des Nations, 1211 Geneva 10, Switzerland

INTRODUCTION

The assessment of potential hazards that may be posed by chemical substances involves, among other things, the critical evaluation of available scientific and technical information pertaining to these substances. The data required for this activity cover a broad range of subject areas and may originate from a variety of national and international sources. Access to the best information in a complete and readily usable form is a prerequisite to sound hazard assessments.

With more than 80 000 chemicals estimated to be in international commerce (OECD, 1986) and with around 1000 chemicals being added each year to this number (Maugh, 1978), there is clearly a significant task ahead of governments to assess both existing chemicals, and chemicals new to commerce, to ensure that their production, manufacture, use and ultimate disposal will not have an adverse impact on human health and the environment. These numbers imply that no one country has the resources to undertake this task alone, and that countries must work together if there is to be any chance of assessing the potential hazards of even the most highly suspect chemical substances. Many countries are at different levels of sophistication in the way in which they conduct their assessments, and the information sources they rely on. As such, international harmonization, international cooperation, and sharing of national experiences and final evaluations is therefore a necessity.

This chapter discusses the activities of the International Register of Potentially Toxic Chemicals (IRPTC) and the relevance of this programme to other international initiatives that are aimed at the preparation of authoritative assessments of chemical hazards. Since information exchange on a global basis is a key component of IRPTC, the chapter also discusses some of the international procedures that have been established to enhance the exchange of information relevant to chemical hazard assessment, and the relationship of IRPTC to these efforts. The chapter also provides details about a recent development called National Registers of Potentially Toxic Chemicals (NRPTCs) which are country-oriented chemical data systems that are compatible with the IRPTC database structure. This concept is emerging as a major direction for IRPTC in an attempt to strengthen the data gathering, data management and hazard assessment capabilities of countries (particularly developing countries), and at the same time trying to push forward the idea of global information exchange among participating countries.

MAIN ACTIVITIES OF IRPTC

The United Nations Conference on the Human Environment held in Stockholm in 1972 recommended that plans be developed for an international registry of data on chemicals (UNCHE, 1972), and subsequently, in 1974, the UNEP Governing Council decided that urgent steps should be taken to establish both a chemicals register and a global network for exchange of the information the register would contain (UNEP, 1974). In 1976, following a further decision by the UNEP Governing Council authorizing the Executive Director to implement IRPTC (UNEP, 1975*a*), and following the development of a strategy and workplan for the project (UNEP, 1975*b*), the IRPTC Programme Activity Centre (PAC) became a reality. The main objectives of IRPTC were to:

(a) facilitate access to existing data on chemicals, thereby making more efficient use of national and international resources available for evaluation;
(b) identify and draw attention to the major gaps in the available information;
(c) help identify potential hazards from chemicals and improve awareness of such hazards; and

(d) provide information about national, regional and global policies, regulatory and other controls of toxic chemicals.

Given these objectives, and the strategy developed by the group of experts in 1975, the central tasks of IRPTC became the development of computerized central files for chemicals, and the operation of a global network for information exchange.

Computerized Database and Chemical Data Profiles

During 1978, a study was undertaken by IRPTC with the assistance of an international group of expert consultants representing a variety of scientific disciplines, to define the attributes (or characteristics) of chemicals that were needed to evaluate the potential hazards posed by these substances to man and the environment. The results of this study, which were published in the same year (IRPTC, 1978), played a significant role in the design of the database files even as they exist today. There are presently 17 'logical' files of data covering everything from the identification and physical/chemical properties of chemicals through to existing regulatory and other controls that have been put into place by countries from representative regions of the world, and by international organizations. In the middle of this spectrum of information are files that describe the behaviour of chemicals released into the environment (e.g. concentrations, environmental fate tests and environmental distribution), and their effects on living systems (e.g. chemobiokinetics, mammalian and special toxicity, and aquatic and terrestrial toxicity). Files dealing with analytical methodology, spills, treatment of poisoning and chemical waste management methods are also included for completeness.

Data quality and data quantity are key issues with the IRPTC database. The objective of IRPTC in this respect has been to provide decision-makers and other users with the most pertinent data available to substantiate their best-informed assessment; benefiting from the assessments made by others is also critical. Providing reliable and detailed data in manageable amounts was the strategy chosen to achieve this objective. Beginning in 1979, IRPTC developed a concise set of instructions for the identification and selection of data being put into the database. The most current instructions (IRPTC, 1985*a*) explicitly define the record structure for each file in such a way that the reader should be able to determine if the study is of adequate quality, and if influencing factors have been appropriately dealt with. The records of similar files (e.g. mammalian, aquatic and terrestrial toxicity files) are consistent in their design while others (e.g. waste and legal files) have unique but clearly defined formats. The structure of all records provides the opportunity to include information such as additional factors that may have influenced the results, author's conclusions where they may be relevant, and the results of authoritative evaluations that may have been undertaken. Each and every record also provides a complete reference to the primary published study as well as the secondary reference when the study has been incorporated into a review document. The role of secondary literature in selecting data for the database is very significant. The IRPTC instructions indicate the need to give certain priority to studies based on their use in international, national and other reviews. By following this requirement, IRPTC is able to incorporate only those studies that have passed the scrutiny of experts, when such a review has taken place, and it ensures that evaluations resulting from these reviews are made available through the database.

The combination of all (or selected) data from each file on a given chemical or class of chemicals is the principal product from the database and is called a Chemical Data Profile. At the present time, fairly complete data profiles exist for about 600 chemicals. The Legal File, which underwent accelerated development to meet the needs of the UNEP Governing Council, contains entries for more than 7000 chemicals. A Register Index is maintained for the database to show a 'profile of the data profiles', and is used to show users quickly the type of information that exists on a chemical as well as identify the gaps that may exist in the profile itself, or in the scientific literature. While most data profiles have been developed using consultants, a number have been developed or updated with the assistance of Contributing Network Partners who are international and national institutions that have agreed to undertake the work as part of their existing efforts to evaluate chemicals.

Dissemination of data from the database takes place in a variety of forms including hardcopy listings of data profiles for users on request, and the publication of specialized reports based on unique files such as the Waste Management File (IRPTC, 1985*b*) and the Legal File (IRPTC, 1986). The data have also been transferred to Canada where the Department of National Health and Welfare is offering, primarily to Canadian users, on-line access to all IRPTC files by subscription.

Copies of the Canadian version of the database have been provided to installations in Thailand, Peru and the People's Republic of China. Finally, on-line access to the Waste Management and Legal Files is being provided through the European Chemicals Data Information Network (ECDIN) of the Commission of the European Communities (CEC) through an agreement for cooperation.

Network for Information Exchange and Related Activities

The development of the IRPTC global network for information exchange, like the work on the computerized database, began at the inception of the Programme Activity Centre in 1976. The network is currently organized much like it was originally conceived. At the heart of the network is the PAC which has responsibility for training, keeping track of, and communicating regularly with the other network participants. Lists of contacts are regularly updated and forwarded through the network to ensure that all participants are aware of each other.

At the country level, the network consists of National Correspondents (NCs) who are government-designated officials having the responsibility to assist the PAC with gathering information on new national regulations or reviews on chemicals, promoting IRPTC and disseminating its publications. There are currently 118 NCs from 110 countries. Also participating at the country level is a group of Designated National Authorities (DNAs), who are responsible for implementing a specific task assigned to IRPTC dealing with the exchange of notifications on national regulatory actions to ban or severely restrict chemicals, as well as other data. Known as the 'London Guidelines for the Exchange of Information on Chemicals in International Trade' (UNEP, 1987), there are currently 78 DNAs from 75 countries involved with this activity.

Contributing Network Partners (CNPs) constitute one of the most important elements of the global network. These national and international institutions include other United Nations bodies, international scientific organizations, national academies of science, specialized research institutions and industrial research centres. They represent the part of the network with the capability to provide data and expertise on chemicals and their effects, to cooperate in data collection and validation, and to provide important links to other institutions. IRPTC relies heavily on these CNPs for assistance with data profile development and as an information resource for many of its other activities.

Associated with this network, IRPTC carries out a number of activities that involve the exchange of information within the formal network, and outside of it. These include the operation of a Query-Response Service, publication of the *IRPTC Bulletin*, sponsorship of training sessions, and the publication of chemical-specific reviews of Soviet scientific literature.

The Query-Response Service, which has been in operation since IRPTC's inception, receives questions daily from national authorities, academia, industry, non-governmental organizations and the public seeking general information but often specific advice on particular chemicals. Staff responding to these queries rely heavily upon the database resources that function within IRPTC as well as a number of external databases to which IRPTC has access. In many cases, however, other sources of information must be accessed and therefore, staff call upon established partners in the network with the required expertise for assistance in developing suitable responses.

The *IRPTC Bulletin* was first published in 1977 as a means for communicating information about the activities of IRPTC, UNEP and other national and international initiatives relating to chemicals. It continues to be published twice a year in four languages (English, French, Russian and Spanish) and provides a useful medium for informing readers about, among other things, new or proposed legislation and regulations for the control of chemicals, international risk evaluation, newly discovered hazards, accident reports and the safe use of chemicals. National Correspondents actively participate in its preparation by providing national information, and in its distribution. In total, the *IRPTC Bulletin* is sent to over 10 000 recipients.

The organization of expert meetings is another forum in which IRPTC facilitates the transfer of information. Since 1980, IRPTC has sponsored, in cooperation with other international programmes, and under the auspices of a joint USSR–UNEP/IRPTC project, almost ten meetings dealing with such matters as the harmonization of approaches to toxicity testing and risk assessment of chemicals, classification systems for chemicals that facilitate their control, and the establishment of regional information exchange networks. In a similar fashion, IRPTC has provided eight training courses and seminars aimed at enhancing the knowledge and skills of participants and their ability to use existing data sources on chemicals in decision-making. Targeted mainly at officials from developing countries, these training activities have covered topics such as preventive toxicology in relation to industrial and household chemicals, international activities concerning chemical control, and the health aspects of pesticide use in agriculture.

Access to scientific data on chemicals that is generated and published in East European countries, as well as access to scientific expertise present in these countries is often difficult. This absence represents a significant gap in chemical knowledge and takes away a major opportunity for global cooperation in dealing with potential chemical hazards. As a result, the State Committee for Science and Technology of the USSR, in cooperation with IRPTC, initiated during 1982, the publication of a series entitled *Scientific Reviews of Soviet Literature on Toxicity and Hazards of Chemicals.* As the title indicates, documents from this series represent comprehensive compilations of Soviet scientific literature on chemicals that are selected on the basis of IRPTC priorities for its database development or from its Query-Response Service. To date, reviews on more than 110 chemicals have been published in this series, and they have been made freely available to those who ask for them.

National Registers of Potentially Toxic Chemicals (NRPTCs)

Following the successful implementation of the computerized IRPTC databases by the PAC, a number of developing countries expressed their interest in creating national registers of potentially toxic chemicals that would be completely compatible with the IRPTC system. Officials from these countries recognized that if they were able to create such systems, data specific to the country (e.g. chemical production and use, concentrations, environmental distribution, legal controls) could be stored and combined with the chemical-specific data that can be found in IRPTC. They felt that the existence of such complementary data systems was essential if sound assessments of chemical hazards and subsequent decisions regarding chemical control at the local level were to be made. IRPTC, believing that the structure of its database was sound, felt that assisting in the development of national systems could ultimately lead to a truly global network for exchange of data between all participants. Therefore, in 1983 with financial support from the Government of the Netherlands, it embarked on a series of pilot projects intended to further define the concept of NRPTCs, to design the database structure such that it would function as a tool to decision-makers at the national level, and to develop guidelines to optimize the use of information that would be contained in the respective databases for hazard assessment.

The first phase pilot project conducted in 1984 involved five participating countries: Colombia, the Gambia, Malaysia, Sri Lanka and Tanzania. Each country was visited by a consultant and introduced to the NRPTC concept. The consultant also gathered information from each country pertaining to the way in which chemical hazards were assessed, and the legal basis for controlling identified hazards. Subsequently, a representative from each country was brought to an intensive 3-week training course in Geneva given by IRPTC staff to teach them about the database structure, the procedures for selecting and extracting data, and the methods for using data from the database, as well as information available from other sources, for assessing hazards. After returning to their countries, the representatives were visited again by the consultant to conduct a follow-up evaluation (Haskoning, 1986). The second phase pilot project was carried out in the same way during 1986 for five more countries: Brazil, China, Indonesia, Thailand and Zambia.

In evaluating the achievements made by participating countries, it was noted that the majority of countries had been able to create national committees among their government agencies responsible for evaluating the hazards posed by chemicals to human health or the environment. Several of these committees met regularly and discussed matters such as the development of national priority lists of chemicals, and the ways to cooperatively develop their national register. Two of the countries had actually taken significant steps toward the implementation of a computerized database; these being Brazil and China (Haskoning, 1986).

A third phase pilot project was undertaken during 1987 and early 1988 in China. Having participated in the second phase, China had accepted the NRPTC concept based on the IRPTC database, and had proceeded very rapidly to the point where it had started data gathering and computerized storage. Realizing that the success of the national register relied on the active involvement and participation of organizations at the provincial level, the National Environmental Protection Agency (NEPA) organized a workshop to conceptualize the organization of the national register in this context. The workshop held in Beijing in April 1988 was highly successful and introduced yet another level into the NRPTC concept—the Provincial Register of Potentially Toxic Chemicals—that is fully compatible with the NRPTC and based on the IRPTC design (Haskoning, 1988).

IRPTC will be continuing with the NRPTC project and is planning on using this initiative as an opportunity to enhance its ability to deliver its data to those who need it in a readily usable form. The conversion of the IRPTC database to a personal computer format is anticipated as a major step towards this goal, and will be very important for most of the participating NRPTCs who do not have access to other types of computer systems. Expanding the number of chemicals covered by the database, and providing more suitable assistance to NRPTCs are also viewed as important targets for the future of the project.

RELATIONSHIP OF IRPTC TO OTHER INTERNATIONAL INITIATIVES

There are a number of international initiatives being undertaken that relate to the preparation of authoritative assessments of chemical hazards, and the exchange of information on chemicals. These include the chemical evaluation activities of the International Programme on Chemical Safety (IPCS), the International Agency for Research on Cancer (IARC), and the Joint FAO/WHO Expert Meeting on Pesticide Residues in Food (JMPR). They also include the efforts of the Organization for Economic Cooperation and Development (OECD) and the Council for Mutual Economic Assistance (CMEA) to enhance information exchange among their member countries and to encourage cooperation.

Evaluation of chemical hazards

The most notable of the international initiatives aimed at the preparation of authoritative assessments of the effects of chemicals on human health and the environment is the IPCS. This programme is a joint venture of the UNEP, the International Labour Organisation (ILO), and the World Health Organization (WHO). In addition to assessments, IPCS undertakes to develop epidemiological, experimental laboratory and risk assessment methodology to produce internationally comparable results, and to provide manpower development in the field of toxicology (IPCS, 1988).

IRPTC has been designated as one of two international IPCS Participating Institutions (the other being IARC), and in this capacity, it contributes data from its data bank and other information sources to the authors of the Environmental Health Criteria documents or Health and Safety Guides. Since the two programmes maintain close working relationships, IRPTC accommodates the needs of IPCS in its own plans to develop chemical data profiles so that activities at the data-gathering step are coordinated. IRPTC also contributes at the later stages of document preparation by providing comments on draft versions and by participating in expert group meetings. Once published, IRPTC makes an effort to incorporate promptly the results of the evaluations into the database to promote dissemination.

In some respects similar to IPCS, IARC undertakes as one of its activities, with the help of international working groups of experts, to prepare and publish critical reviews and evaluations of evidence on the carcinogenicity of a wide range of agents to which humans may be exposed (IARC, 1988). Appearing in the series entitled *IARC Monographs on the Evaluation of Carcinogenic Risks to Humans,* these evaluations are systematically incorporated into the IRPTC files concerning genotoxicity. Furthermore, the supporting studies are extracted and entered into the database as full records. To date, evaluations for over 600 chemicals that have been produced by IARC now appear in the IRPTC database.

The Joint FAO/WHO Expert Meetings on Pesticide Residues in Food publish monographs that summarize the safety data on those pesticide residues for which there are sufficient data for the expert group to make decisions regarding acceptable daily intake (FAO/WHO, 1988). Since this evaluation is of considerable importance to a comprehensive assessment of hazard, IRPTC is working to incorporate the results in the IRPTC database. Preliminary investigations are also being made to determine if IRPTC may be able to assist in the management of data considered during the expert meetings and during the data-gathering step.

Information Exchange

Over the years OECD has played an important role in trying to achieve harmonization and promote cooperation among its member states, and with other international organizations including those in the United Nations system and the European Economic Community. The scope of their efforts has covered matters such as harmonizing the approaches to the control of chemicals, harmonizing the test methods available to generate data useful for assessing hazards of chemicals, setting the principles of Good Laboratory Practice to ensure the quality of test data, and harmonizing data requirements and assessment practices. Information exchange has also been an important topic for the OECD Chemicals Group, and has led to recommendations concerning the exchange of information related to the export of hazardous chemicals, and the exchange of confidential data that may be submitted to member states. They have also implemented exchange procedures for data of a more general nature through activities such as the Complementary Information Exchange Procedure (CIEP), the Chemicals Group Forum, and the OECD Switchboard. The work related to export notification is carried out in close cooperation with IRPTC/UNEP. Similarly, IRPTC participates in the OECD Switchboard. Recent developments by OECD with EXICHEM have led to another opportunity for cooperation between OECD and IRPTC. At present, IRPTC provides information on the existence of critical reviews on chemicals that are identified by its network contacts.

Under the auspices of the joint USSR–UNEP/IRPTC project, member countries of the CMEA and its secretariat have embarked on an activity aimed at establishing a regional information exchange network on chemicals. The IRPTC database is being considered as the basis for this network, and if adopted could lead to another mechanism for the exchange of information with East European countries (IRPTC, 1989).

The computerized listing of Chemicals Currently being Tested for Toxic Effects (CCTTE) is a joint venture by IPCS and IRPTC with the purpose of encouraging information exchange between research institutions which undertake long-term and expensive toxicological studies. In doing so, CCTTE should minimize the risk of different institutions expending large resources doing essentially the same work; it should also encourage cooperative efforts. A second part to CCTTE deals with the identification of national efforts that are underway or have recently been completed, to prepare comprehensive critical reviews on chemicals. IRPTC has made the necessary provisions in its library and documentation system such that its acts as an international repository for such reviews. CCTTE is published once a year in hard-copy.

CONCLUSIONS

In conclusion, it can be said that IRPTC and other international programmes and organizations such as IPCS, IARC, FAO/WHO, OECD and CMEA have been making significant progress in preparing authoritative assessments on chemicals, and in sharing this information with others. IRPTC, with its carefully designed database structure, and with its efforts to transfer its data management capability and its data to others, provides a sound model for national and regional data systems, and brings consistency to information exchange procedures. It is contributing to the preparation of hazard assessments by providing reliable and detailed data, and it is disseminating the results widely. Furthermore, IRPTC is providing, in cooperation with the USSR, access to data sources that were not previously readily available. Therefore, IRPTC is serving as an essential international tool for the assessment of chemical hazards and for the exchange of information on chemicals.

REFERENCES

FAO/WHO (1988). Pesticide residues in food—1987; Evaluations 1987 Part II—Toxicology. FAO Plant Production and Protection Paper 86/2, Food and Agriculture Organization, Rome, Italy.

Haskoning (1986). Assistance to selected developing countries in establishing national information systems on potentially toxic chemicals—Phase II final report. Unpublished contract report, Berg en Dalseweg, the Netherlands.

Haskoning (1988). Establishment of a national information system on potentially toxic chemicals—Evaluation. Unpublished contract report of the Workshop, Beijing, 18–29 April 1988, Berg en Dalseweg, the Netherlands.

IARC (1988). *IARC Monographs on the Evaluation of Carcinogenic Risks to Humans,* vol. 43. World Health Organization, Lyon, France, pp. 15–32.

IPCS (1988). Vanadium. Environmental Health Criteria 81, World Health Organization, Geneva, Switzerland.

IRPTC (1978). Attributes for the Chemical Data Register of the International Register of Potentially Toxic Chemicals, Register Attribute Series, No. 1. International Register of Potentially Toxic Chemicals, UNEP, Geneva, Switzerland.

IRPTC (1985*a*). *International Register of Potentially Toxic Chemicals,* Part A. International Register of Potentially Toxic Chemicals, UNEP, Geneva, Switzerland.

IRPTC (1985*b*). Treatment and disposal methods for waste chemicals, Data Profile Series No. 5, International Register of Potentially Toxic Chemicals, UNEP, Geneva, Switzerland.

IRPTC (1986). IRPTC Legal File 1986, Vols I and II, IRPTC Data Profile Series No. 7. International Register of Potentially Toxic Chemicals, UNEP, Geneva, Switzerland.

IRPTC (1989). Expert Consultation on a Regional Information Exchange Network for Chemicals, Moscow, 18–21 October 1988.

Maugh II, T.H. (1978). Chemicals: how many are there? Science, **199,** 162.

OECD (1986). *Existing Chemicals—Systematic Investigation, Priority Setting and Chemical Reviews.* Organization for Economic Cooperation and Development, Paris, France.

UNCHE (1972). The United Nations Conference on the Human Environment, Stockholm, 5–16 June 1972, Recommendation 74(e).

UNEP (1974). Governing Council of the United Nations Environment Programme, second session, Decision 8 A (II), section I, paragraph 1 (k).

UNEP (1975*a*). Governing Council of the United Nations Environment Programme, third session, Decision 29 (III), paragraph 8.

UNEP (1975*b*). Report of the UNEP Expert Workshop on an International Register of Potentially Toxic Chemicals, Bilthoven, Netherlands, 6–11 January, Document UNEP/WG.1/4.

UNEP (1987). Governing Council of the United Nations Environment Programme, fourteenth session, Decision 14/27.

4

Environmental Hazard Assessment of Anthropogenic Chemicals

WALTER KLÖPFFER

Battelle-Institut e.V., Abteilung Chemische Analytik, Am Römerhof 35, D-6000 Frankfurt am Main 90, FRG

INTRODUCTION AND GENERAL CRITERIA

Reasons for Assessment Work

The purpose of environmental assessment of chemicals is to recognize the potential hazard posed by chemicals before actual damage occurs. It is thus possible to take decisions and preventative actions. The main reasons for the importance of this are:

(a) The responsibility we have for the generations which will come after us.
(b) The fact that the existence of mankind depends on the biosphere but not the other way round. The 'cultural sphere' (Bretschko, 1983) or 'technosphere' (Frische *et al.*, 1982) is of very recent origin compared with the much longer lasting biosphere and the energy/matter basis below, and strongly depends on the state of its substrate. Drastic changes of the biosphere may therefore destroy the basis of our existence.
(c) Chemical legislation as an important part of environmental protection is the immediate incentive for assessing the potential of chemicals to damage the biosphere (or the 'environment' as the counterpart of the technosphere).

Chemical legislation has been established, for example, in the following countries: Japan (1973), Sweden (1973), USA (1976) ('TSCA'), France (1977), EC (1979) ('6th Amendment'), FRG (1980) ('ChemG'), Switzerland (1983) and Austria (1987).

The goal of the chemical laws is to protect human health and the environment from adverse effects of chemicals. Assessment work is particularly needed in two areas:

(1) Assessment of 'existing chemicals', i.e. to sort out the probably small percentage of dangerous substances from roughly 100 000 chemicals which are currently manufactured (OECD Chemicals Group, 1984; CRCS, 1985; OECD, 1986; GDCh, 1988; Greim & Sterzl, 1988; Weiß *et al.*, 1988).
(2) Assessment of 'new chemicals' (Schmidt-Bleek & Wagenknecht, 1979; UBA, 1980; Brown, 1982; Schmidt-Bleek *et al.*, 1982; Hushon *et al.*, 1983; Rippen *et al.*, 1985).

Environmental Hazard Criteria of Chemicals

In assessing the environmental hazard of a chemical it is useful to start with some general criteria (Frische *et al.*, 1979, 1982):

Persistence means the longevity of a compound in the environment and is caused by the absence or inefficiency of sinks for the substance.

Mobility is the dispersion tendency of a substance. Mobile substances can be distributed worldwide, if they live long enough.

Amount is usually defined as the annual input into the environment and may range from about 0·1% to 100% of the annual production of a chemical.

Noxious effects of a chemical may be directed against man (toxicity) or against animals, plants or ecosystems (ecotoxicity) or against non-biological targets, from statues and cathedrals up to the ozone layer.

Accumulation is closely related to persistence and causes high local concentrations, e.g. in sediments and fat tissues of animals.

These criteria are, in general, too broad to be used directly for a detailed hazard assessment, but they are useful for qualitative descriptions and for the general understanding of a chemical's environmental behaviour.

WHAT IS ENVIRONMENTAL HAZARD?

There is considerable confusion about the meaning of the term 'hazard', even within the OECD Chemicals Program (Frick, 1982; OECD Hazard Assessment Project, 1982*a–d*; OECD, 1982; Richardson, 1986, 1989). A useful, albeit not very precise definition has been given by the OECD Hazard Assessment Project, (1982*d*): 'The hazard of a chemical is a function of two broad considerations, the potential of the chemical to harm biological systems (or damage other systems) and its potential for exposure such that the harm or damage can occur.' According to this definition, the hazard is determined by exposure and effects, both biological (toxic/ecotoxic) and non-biological ones.

The term 'hazard' is also often confused with 'risk'. Risk assessment, however, contains a quantitative probability estimate, which is very difficult to obtain for the case of general environmental (as opposed to direct human) exposure. I therefore agree with Richardson (1989) that risk assessment goes one step beyond the hazard estimate (Fig. 1).

As a visualisation of the hazard definition, it is convenient to write the following equation (Klöpffer & Rippen, 1983; OECD, 1982):

$$\text{Hazard} = \text{Exposure} \times \text{Effects}$$

The multiplication sign (×) indicates that there should be no hazard if there is either no exposure or no (noxious) effect. This seems to be widely accepted, but in a strict sense contradicts the basic postulate of environmental hazard assessment that 'it is scientifically impossible to prove that a substance will *not* pose a threat' (OECD Hazard Assessment Project, 1982*a*).

As already shown in Fig. 1, most assessment systems involve separate exposure and effect assessments. The accuracy of the results

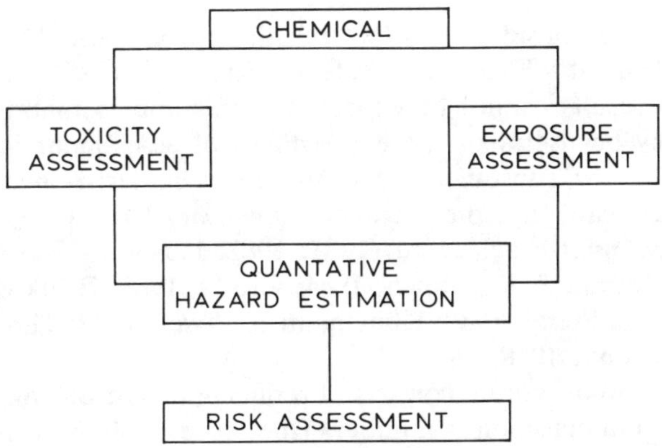

FIG. 1. Scheme of hazard estimation and risk assessment after Richardson (1989).

depend, of course, on the amount and quality of data available. Chemical legislation provides for tiered (mostly three-level) testing systems, which depend on the annual production volume of the chemical (EC, 1979; FRG, 1980; OECD Hazard Assessment Group, 1982*a–d*; Rippen *et al.*, 1985). For exposure analysis, at least some useful data are available at the zero-level (or minimum pre-marketing (MPD) level) of the testing system for new chemicals (Hushon & Clerman, 1981; OECD Hazard Assessment Project, 1982*a*; Klöpffer *et al.*, 1982; Hushon *et al.*, 1983). The situation is very poor, however, for effect data (OECD Hazard Assessment Project, 1982*b,c*; Klöpffer & Rippen, 1983).

According to Rudolph & Boje (1987), the ecotoxic effects should be recorded at the different levels of organisation: from the cells, organs, organisms, etc. up to ecosystems and the biosphere as a whole. This information cannot be provided by any one testing system, however. Only a few single-species tests are available at the zero-level; for levels 1 and 2 the situation is somewhat better, but still far from satisfactory.

For some existing chemicals that have long been suspected of being hazardous, many data are available, although only few, at the ecosystem level. Data on non-biological systems are available if by chance adverse effects have been detected or correctly predicted. Nearly nothing is known about the effects of transformation products formed in the environment.

PRINCIPLES OF SCORING SYSTEMS

Scoring may be considered as the 'zeroth approximation' to solve the equation: Hazard = Exposure × Effects (Klöpffer, 1985). As in physics, precise results cannot be expected in this approximation! Scoring systems may be useful in the early stages of assessment in chemical notification (new substances) and in screening large inventories of existing chemicals, e.g. the European inventory EINECS (Lüderitz *et al.*, 1980; Schmidt-Bleek *et al.*, 1981, 1982; Hushon, 1983; Jouany *et al.*, 1983; Hushon & Kornreich, 1984; CRCS, 1985; Brink & Walker, 1987; Greim & Sterzl, 1988; Könemann & Visser, 1988; Timmer *et al.*, 1988; Weiß *et al.*, 1988).

The principle of scoring consists of reducing test results or calculated properties to dimensionless numbers (the scores), often in three steps: high (+3)—medium (+2)—low (+1). The scores are added up to a total score which approximates the hazard. In more elaborate scoring systems this may be done separately for the compartments water, air, soil and sediment after some compartmentalisation exercise (Frische *et al.*, 1979; Schmidt-Bleek *et al.*, 1981; Klöpffer *et al.*, 1982; Brown, 1982; Mackay *et al.*, 1983; Weiß *et al.*, 1988).

Another feature is the weighting factors for individual scores, which may (Frische *et al.*, 1979, 1982) or may not (Könemann & Visser, 1988) be introduced in order to fit the results to a ranking obtained by expert judgement. Figure 2 shows the results obtained with weighted scores for several chemicals, ranging from relatively harmless to hazardous. In this case, the sum of weighted scores in a formalised scoring scheme is compared with the results obtained in expert sessions after intense discussion and using all information available at that time. For the scoring exercise, much less information has been used in order to simulate the situation of initial assessment of new chemicals (Frische *et al.*, 1979).

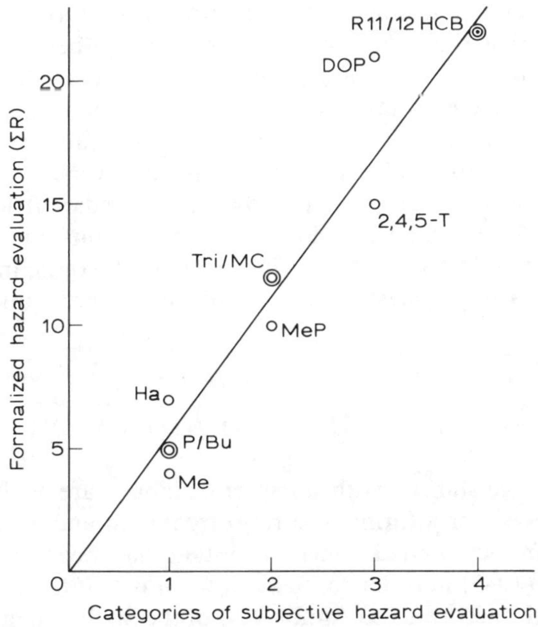

FIG. 2. Environmental hazard assessment of chemicals after Frische *et al.* (1979, 1982).

There are too many scoring systems in existence to discuss any details here; Judy Hushon summarised 33 systems at the Copenhagen Symposium in 1982 (Christiansen *et al.*, 1983; Hushon, 1983). In my view, a few general requirements should be met by scoring (and other simple hazard ranking) systems in order for them to be useful (Klöpffer, 1985). The systems must be:

(a) in accordance with experience on well-known existing chemicals (calibration!);
(b) compatible with assessment at higher levels;
(c) conservative (no false positives!); and
(d) transparent for all users.

First of all, the scoring system has to be in accordance with experience gained from well-known environmental chemicals. The calibration process is not unproblematic, since hazard assessment is not performed in the same way by different expert groups, agencies or individuals.

Second, the system should be compatible with the assessment at higher levels of the testing and evaluation procedure. This demand cannot mean that the results must be identical (otherwise the higher levels would not be necessary!), but it means that the assessment procedures at all levels must be conceived in the same spirit.

Due to the limited database at the zero-level, the results obtained should be more conservative than at the higher levels.

Last, but not least, the system should be transparent for all users, including the notifier; he should know what results the agency will obtain out of the data submitted. This last point favours, in my view, a formalised ranking at least as a first step of hazard ranking of new substances.

ADVANCED HAZARD ASSESSMENT

If only a few substances with a better database are to be assessed, there is less need for a formal (scoring) treatment and a 'case-by-case' procedure can be applied. Such a system has been developed by Rippen *et al.* (1985) in order to make optimal use of the data provided by levels 1 and 2 according to Annex VIII of the Sixth Amendment (EC, 1979). Further advanced approaches have been presented by Weber & Barben (1984) and recently by Landner (1988).

The exposure analysis is not formalised in this approach and can be done, starting from the use and estimated release pattern, using appropriate multimedia models (Rippen *et al.*, 1983, 1984; Mackay *et al.*, 1983; Frische *et al.*, 1984). If one compartment or medium, e.g. water, is strongly preferred, a more sophisticated model can be used. In the frequent case of surface water as the main receiving compartment, EXAMS may be used, or an easy-to-use model for IBM-PC, which has recently been developed for Umweltbundesamt, Berlin (Frank & Klöpffer, 1989).

The hazard assessment system of Rippen and co-workers consists essentially of series of comprehensive and coherent checklists for each compartment and for the essential pathways by which man can be exposed to the chemical (indirect human exposure). Figure 3 shows, as an example, the list for the compartment 'surface water'. The starting point is the discharge pattern, which must be known for air (L), soil (B) and water (wastewater and surface water). Estimations have to be made with regard to the amounts and potential concentrations in both subcompartments (wastewater still belongs, strictly speaking, to the technosphere; Frische *et al.*, 1979, 1982). Questions to be answered concern possible transformation products, transfer into sewage sludge and soil, and volatility both from the sewage treatment plant and from surface water. Concentrations then have to be estimated for sea water and fresh water. Next the data are compared with suitable data on the ecotoxicity of the substance. If the answer to the question on transformation (degradation) in water is 'no' the result 'persistent in water' is given. If the safety margin is >1000, the result reads: 'no endangering of aquatic organisms perceptible', but if the safety margin is <1000 and is confirmed by additional tests, the result reads: 'endangering of aquatic organisms is possible'.

EXTREMELY TOXIC AND PERSISTENT CHEMICALS

The hazard assessment based on the principle Hazard = Exposure × Effects in general underrates two limiting cases: (a) very high toxicity with small or not recognisable exposure (e.g. confinement in closed systems) and (b) exposure without evident noxious effects.

The first case has been taken into account by Weiß *et al.* (1988) in a scoring system for existing chemicals by introducing an 'extra black box' for highly toxic chemicals with low production volume. This

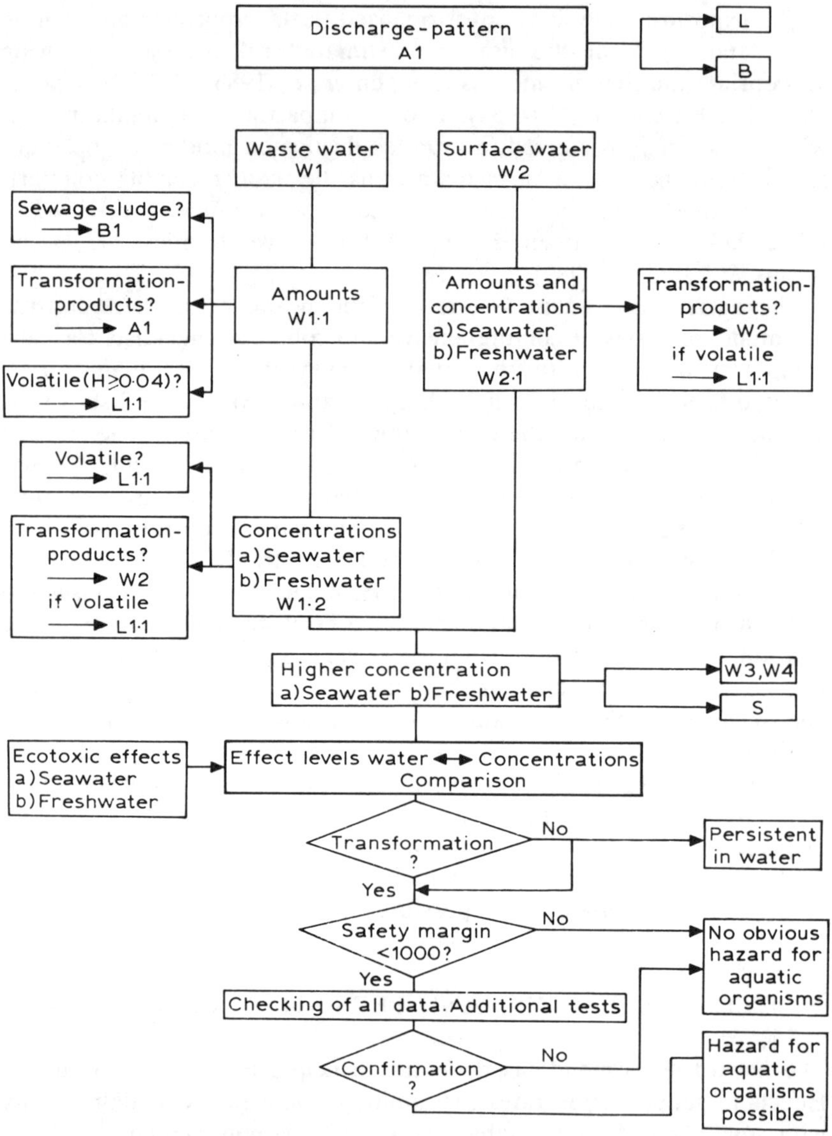

FIG. 3. Hazard assessment system by Rippen *et al.* (1985): check list for the subsystem 'Surface Water'. L, B, and S point to similar check lists for the compartments air, soil, water and sediment (ecosystems); W3, W4 point to the human exposure check lists (exposure route via water).

should give a warning sign of special danger, so that these compounds are not overlooked in the selection procedure.

The second weak point is due to persistent chemicals, which in general have low acute toxicity (Stephenson, 1977; Frische *et al.*, 1979, 1982; Klöpffer, 1989). The first recognition of persistent chemicals in waters goes back to the 1950s. The Report of the Committee on Synthetic Detergents by the UK Ministry of Housing and Local Government (1956) contains impressive documentation (e.g. a picture of the sewage treatment plant of Motherwell in May 1954) of foam due to persistent or 'refractory' tensides contained in the detergents of that time. Since then these so-called hard tensides have been banned and replaced by degradable anionic tensides (LAS).

The first recognition of the general importance of persistence is due to Stephenson (1977):

> 'The question of the importance attached to persistence as an effect in itself is a very difficult one. On the face of it there appears little reason to be concerned about a material which, even though present in the environment, is not causing any detectable damage. On the other hand, persistent materials, because of this property, will accumulate in the environment for as long as they are released. Since the environment is not effective at cleansing itself of these materials, they will remain for indefinite periods. . . . The problem could become entirely out of control and it would be extremely difficult if not impossible to do anything about it.'

The main point in this statement is the fact that persistent chemicals cannot be removed from the environment, once a noxious effect has been detected.

A further big step in understanding environmental chemicals is the recognition that apparent toxicity and environmental hazard may often be unrelated as stated by Gebauer (1986): 'It's a dilemma—the more toxic chemicals generally are those which are more tolerable to the environment, while the low toxicity materials may often have more environmental problems.' This is certainly true in general and insofar as acute toxicity is concerned. Acute toxicity is not, of course, a major problem in the environment, due to the typically very low concentrations, but it is a problem for safe handling of the chemicals in the technosphere.

Important examples of persistent chemicals are as follows (Klöpffer,

1989):

tetrapropylene benzenesulphonate (TPBS)
perchloro- and chlorofluorohydrocarbons
1,1,1-trichloroethane ('methylchloroform')
polychlorinated biphenyls (PCBs)
DDT/DDD/DDE
di(2-ethylhexyl)phthalate (DEHP)
silicones (?)
polychlorinated dibenzodioxines (PCDD)
and polychlorinated dibenzofurans (PCDF)

These well-known persistent chemicals or 'xenobiotica' (Korte *et al.*, 1987) have been detected in many environmental samples. The first example, TPBS, is the tenside which caused the foam documented in the British Report of 1956. It has now been replaced by more easily degradable substances. Perchlorinated compounds, and especially the chlorofluorohydrocarbons, are good examples of compounds formerly considered to be absolutely safe. The ozone-degrading activity in the stratosphere is only possible due to their long lifetime in the troposphere. Methylchloroform is less persistent than the chlorofluorohydrocarbons but is still long-lived enough to contribute to the total chlorine input into the stratosphere.

The polychlorinated biphenyls (PCBs) have been used in transformers, capacitors, in mining and many other uses on account of their excellent chemical and physical properties. Although their use has been confined to so-called closed systems, PCBs have been found in sediments, soils, and in fat tissues of animals and man. It is clearly a persistent and highly accumulating mixture. The production of PCBs has recently been suspended in the FRG.

The 'DDT family' of closely related compounds is the oldest known group of persistent chemicals. DDE is the main transformation product of DDT and is even more stable than the mother compound. DDD also seems to be more stable than DDT.

DEHP is a well-known plasticiser and widely distributed in sediments, sewage sludge, etc. It is reported to be biodegradable, but evidently this does not prevent its distribution in the environment. It is not degraded in anaerobic sediments and is therefore protected in this environment.

Whereas the persistence of silicones in the environment is still a controversial issue (Klöpffer, 1989), there can be no doubt that the polychlorinated dibenzodioxines (PCDDs) and dibenzofurans (PCDFs) are persistent chemicals. We are confronted here with the very rare case of *extremely toxic substances that are also persistent.* This is the worst possible combination of properties. PCDDs and PCDFs are produced only in very small quantities for research purposes, but they are also accidentally formed in several processes as impurities within the technosphere and are emitted into the environment. The environmental exposure, therefore, is relatively small. In order not to underestimate the hazard of such chemicals, the 'hazard equation' should be written in the following form:

$$\text{Hazard} = \text{Exposure} + \text{Effects}$$

This is, of course, a formal expression with little practical consequence. It should, however, express symbolically that chemicals with either very high exposure potential without known effects, or extreme toxicity coupled with very low or not recognizable exposure potential should be considered as an environmental hazard.

I would like to close this chapter by pointing out two of the many unsolved problems (Staab, 1985; Henschler, 1988):

(a) the provision of safe substitution products for banned chemicals; and
(b) achieving a balance between benefits for human health and protection of the environment.

The first addresses the question of substitution products. It is not sufficient to ban a known hazardous substance without thinking about substitution products or methods, which may also pose environmental problems of a similar or different nature. The second problem is mostly discussed with regard to DDT (Staab, 1985; Henschler, 1988), which had great beneficial effects in combating malaria and for the discovery of which Paul Müller was honoured with the Nobel Prize in Medicine in 1948. Environmentally more acceptable substitution products seem to be more expensive, so that Third World countries cannot afford them. The only solution I can suggest is that the more expensive insecticides are provided by means of development aid. This would also have a beneficial effect for the industrialised countries in that it would reduce the 'import' of DDT via the atmosphere and products from these countries.

REFERENCES

Austria (1987). 326. Bundesgesetz vom 25. Juni 1987 über den Schutz des Menschen und der Umwelt vor Chemikalien (Chemikaliengesetz—ChemG). Bundesgesetzbl. 23 Juli 1987.

Bretschko, G. (1983). Donaunutzung und Donauforschung. In *Internationale Arbeitsgemeinschaft Donauforschung der Societas Internationalis Limnologiae,* 23, Arbeitstagung Wien 13, bis 17, September 1982, Verlag der Österr. Akademie der Wißenschaften, Wien.

Brink, R. H. & Walker, J. D. (1987). Interim Report: TSCA Interagency Testing Committee. Sixth Scoring Exercise. EPA Contract No. 68-02-4251. Dynamic Corp., Maryland.

Brown, S. L. (1982). Steps toward environmental hazard assessment of new chemicals: An independent review. Report of SRI International to Umweltbundesamt Berlin No 82-106 02-025, Berlin.

Christiansen, K., Koch, B. & Bro-Rasmussen, F. (eds) (1983). *Chemicals in the Environment,* Proceedings Int. Symp. Lyngby-Copenhagen, 18–20 October 1982. Technical University of Denmark, Lyngby.

CRCS Inc. (1985). Technical Support to the TSCA Interagency Testing Committee; in Collaboration with Dynamac Co., Rockville, MD, work performed under EPA Contract No. 68-01-6650, September.

EC (1979). Council Directive 79—831/EEC Amending for the Sixth Time Directive 67/548/EEC on the Classification, Packaging and Labelling of Dangerous Substances. Off. J. L 259/10 15 October.

France (1977). Loi No. 77.771 du 12 Juillet 1977 sur le contrôle des produits chimiques.

Frank, R. & Klöpffer, W. (1989). A convenient model and program for the assessment of abiotic degradation of chemicals in natural waters. *Ecotox. Environ. Safety,* **17,** 323–32.

Frauerwieser, G., Mahin Foroutan-Rad, M., Luger, G., Klöpffer, W. & Maly, K. H. (1986). Grundlagen für die Ausgestaltung des österreichischen Chemikalienrechts und dessen Vollziehung. Österreichisches Bundesinstitut für Gesundheitswesen.

FRG (1980). Gesetz zum Schutz vor gefährlichen Stoffen (Chemikaliengesetz—ChemG) vom 16 September 1980. Bundesgesetzbl. Tl. I (1980) 1718.

Frick, G. W. (ed.) (1982). *Environmental Glossary* (2nd Edn). Governmental Institutes Inc., Rockville, MD, USA.

Frische, R., Klöpffer, W. & Schönborn, W. (1979). Bewertung von organisch-chemischen Stoffen und Produkten in Bezug auf ihr Umweltverhalten—chemische, biologische und wirtschaftliche Aspekte, 1 und 2 Teil. Bericht des Battelle-Instituts e.V., Frankfurt am Main, an das Umweltbundesamt, Berlin, Forschungsbericht 101 04 009/03.

Frische, R., Esser, G., Schönborn, W. & Klöpffer, W. (1982). Criteria for assessing the environmental behavior of chemicals: Selection and preliminary quantification. *Ecotox. Environ. Safety,* **6,** 283–93.

Frische, R., Klöpffer, W., Rippen, G. & Günther, K.-O. (1984). The environmental model segment approach for estimating potential environmental concentrations—Part I: The model. *Ecotox. Environ. Safety,* **8,** 352–62.

Gebauer, Werner (1986). Portrait of an industry leader. *Farm Chemicals International,* **1**(1), 16A–16B.

GDCh (1988). *Altstoffbeurteilung—Ein Beitrag zur Verbesserung der Umwelt* (2nd Edn). Beratergremium für umweltrelevante Altstoffe (BUA) (Herausg.) Frankfurt am Main.

Greim, H. & Sterzl, H. (1988). 100 000 Alte Stoffe im Verkehr?! Zum Lösungsansatz der Naturwissenschaften. *Entsorgungspraxis Spezial* No. 3, 130–5.

Henschler, D. (1988). Verhältnismäßigkeit im Umweltschutz. *Nachr. Chem. Tech. Lab.,* **36,** 985–8.

Hushon, J. (1983). Survey of scoring systems for hazard assessment. In *Chemicals in the Environment,* Proceedings Int. Symp. Lyngby-Copenhagen 18–20 October 1982, eds K. Christiansen, B. Koch & F. Bro-Rasmussen. Technical University of Denmark, Lyngby, pp. 435–50.

Hushon, J. & Clerman, R. J. (1981). Estimation of exposure to hazardous chemicals. In *Hazard Assessment of Chemicals—Current Developments,* Vol. 1, eds J. Saxena & F. Fisher. Academic Press, New York, pp. 323–88.

Hushon, J. M. & Kornreich, M. R. (1984). Scoring systems for hazard assessment. In *Hazard Assessment of Chemicals: Current Developments,* Vol. 3, ed. J. Saxena. Academic Press, New York, pp. 63–109.

Hushon, J. M., Klein, A. W., Strachan, W. J. M. & Schmidt-Bleek, F. (1983). Use of OECD premarket data in environmental exposure analysis for new chemicals. *Chemosphere,* **12,** 887.

Japan (1973). Chemical Substances Control Law, Law No. 117.

Jouany, J.-M., Vaillant, M., Blarez, B., Cabridenc, R., Ducloux, R. & Schmitt, S. (1983). Une méthode qualitative d'appréciation des dossiers en écotoxicologie: cas des substances chimiques. *Sci. Vét. Méd. Comp.,* **85,** 5–23.

King, N. J. & Hinchcliffe, P. R. (1981). The assessment of environmental hazards posed by chemicals. *Chem. Ind.,* **5,** 585–7.

Klöpffer, W. (1985). Minimum criteria for hazard ranking. Paper presented at the International Workshop on Environmental Hazard Assessment of Chemicals. Schmallenberg-Grafschaft, FRG, 26/27 September.

Klöpffer, W. (1989). Persistenz und Abbaubarkeit in der Beurteilung des Umweltverhaltens anthropogener Chemikalien. *UWSF-Z. Umweltchem. Ökotox.,* **1**(2), 43–51.

Klöpffer, W. & Rippen, G. (1983). Assessment of environmental hazard of substances and mixtures. Integration of the Final Reports of the three Working Parties within the OECD Hazard Assessment Project. Report No 106 04 014, Battelle-Institut e.V., Frankfurt am Main, to Umweltbundesamt, Berlin, January.

Klöpffer, W., Rippen, G. & Frische, R. (1982). Physicochemical properties as useful tools for predicting the environmental fate of organic chemicals. *Ecotex. Environ. Safety,* **6,** 294–301.

Könemann, H. & Visser, R. (1988). Selection of chemicals with high hazard potential: Part 1: WMS scoring system. *Chemosphere,* **17,** 1905–19.

Korte, F., Bahadir, M., Klein, W., Lay, J. P. & Parlar, H. (1987). *Lehrbuch der Ökologischen Chemie.* Thieme, Stuttgart.

Landner, L. (1988). Advanced hazard assessment: Hazardous chemicals in the environment—Some new approaches to advanced assessment. *Ambio,* **17,** 360–6.

Lüderitz, P., Koch, R., Dobberkan, H.-J., Strobel, K. & Hönig, H. (1980). Kriterien zur Begründung einer Rangfolge für eine umfassende hygienisch-toxikologische Untersuchung potentieller Verunreinigungen der Gewässer zur Festlegung von hygienischen Grenzwerte. *Acta hydrochim. hydrobiol.,* **8,** 631–4.

Mackay, D., Paterson, S. & Joy, M. (1983). Fate of chemicals in the environment: Compartmental and multimedia models for prediction. *ACS Symposia Series,* **225,** 175–96.

Ministry of Housing and Local Government (1956). Report of the Committee on Synthetic Detergents. Her Majesty's Stationery Office, London.

OECD (1982). *Chemical Control Legislation—An International Glossary of Key Terms.* Organisation for Economic Cooperation and Development, Paris.

OECD (1986). *Existing Chemicals—Systematic Interpretation Priority Setting and Chemical Reviews.* Organisation for Economic Cooperation and Development, Paris.

OECD Chemicals Group (1984). Chemicals on which data are currently inadequate: Selection criteria for health and environment purposes. OECD Existing Chemicals Programme. Final Report of Expert Groups III and IV, Vols 1 & 2. Istituto Superiori di Sanità, Rome and Umweltbundesamt, Berlin.

OECD Hazard Assessment Project (1982*a*). Step Systems Group: Final Report of the Working Party on Exposure Analysis, Umweltbundesamt, Berlin, March.

OECD Hazard Assessment Project (1982*b*). Step Systems Group: Final Report of the Working Party on Natural Environment Effects, February.

OECD Hazard Assessment Project (1982*c*). Step Systems Group: Final Report of the Working Party on Health Effects. Washington, DC, February.

OECD Hazard Assessment Project (1982*d*). Step Systems Group: Final Report. Stockholm, February.

Der Rat von Sachverständigen für Umweltfragen (1987). *Umweltgutachten 1987.* Verlag W. Kohlhammer, Stuttgart.

Richardson, M. L. (ed.) (1986). *Toxic Hazard Assessment of Chemicals.* Royal Society of Chemists, London.

Richardson, M. L. (1989). European Conference on Chemistry and the Environment. Working Party of the Federation of European Chemical Societies (FECS. Epilogue. Royal Society of Chemists, London. Reprinted in *ÖCHEMZ,* **90,** 17–19.

Rippen, G., Frische, R., Klöpffer, W. & Günther, K.-O. (1983). The environmental model segment approach for estimating potential environmental concentrations. In *Chemicals in the Environment,* Proceedings Int.

Symp. Lyngby-Copenhagen, 18–20 October 1982, eds K. Christiansen, B. Koch & F. Bro-Rasmussen. Technical University of Denmark, Lyngby, pp. 87–97.

Rippen, G., Frische, R., Klöpffer, W. & Günther, K.-O. (1984). The Environmental Model Segment Approach for Estimating Potential Environmental Concentrations—Part II: Application of the Model to *p*-Dichlorobenzene and Trichloroethene. *Ecotox. Environ. Safety,* **8,** 363–77.

Rippen, G., Dumpert, K., Klöpffer, W. & Schönborn, W. (1985). Systemanalytische Erprobung von Verfahren zur Bewertung der Umweltgefährlichkeit von neuen Stoffen im Hinblick auf das Chemikaliengesetz. Bericht des Battelle-Instituts e.V., Frankfurt am Main, für das Umweltbundesamt, Berlin, Forschungsvorhaben No. 10604020.

Rudolph, P. & Boje, R. (1987). Ökotoxikologie nach dem Chemikaliengesetz. Grundlagen für die ökotoxikologische Bewertung von Umweltchemikalien. In *Handbuch der Umweltchemikalien. Loseblattsamlung* (2nd Edn), Vol. 1, ed. G. Rippen. ecomed, Landsberg am Lech, FRG, pp. 5–105.

Staab, H. A. (1985). Das Bild der Chemie im Wandel der letzten hundert Jahre. Festvortrag zum 100 jährigen Bestehen der Berufsgenossenschaft der chemischen Industrie 5, June 1985. Sonderdruck der Berufsgenossenschaft der chemischen Industrie, Heidelberg.

Schmidt-Bleek, F. & Wagenknecht, P. (1979). Umweltchemikalien—Das Problem der Umweltchemikalien vor der Verabschiedung eines Chemikaliengetzes in der Bundesrepublik Deutschland. *Chemosphere,* **8,** 583–721.

Schmidt-Bleek, F., Haberland, W. & Klein, A. (1981). Approaches to environmental hazard assessment based on European Directive 79/831/EEC, Annex VII. EEC/EPA Workshop on Hazard Assessment, Rome (Manuscript).

Schmidt-Bleek, F., Haberland, W., Klein, A. W. & Caroli, S. (1982). Steps towards environmental hazard assessment of new chemicals (including a hazard ranking scheme, based on directive 79/831/EEC). *Chemosphere,* **11,** 383–415.

Stephenson, M. S. (1977). An approach to the identification of organic compounds hazardous to the environment and human health. *Ecotox. Environ. Safety* **1,** 39–48.

Sweden (1973). Law on Products Hazardous to the Health and the Environment of April 27, 1973, SFS 1973:329.

Switzerland (1983). Bundesgesetz vom 7. Oktober 1983 über den Umweltschutz (Umweltschutzgesetz—USG).

Timmer, M., Könemann, H. & Visser, R. (1988). Selection of chemicals with high hazard potential: Part 2: Application and results of the WMS scoring system. *Chemosphere,* **17,** 1921–34.

UBA (1980). *Umweltchemikalien—Prüfung und Bewertung von Stoffen auf ihre Umweltgefährlichkeit im Sinne des neuen Chemikaliengesetzes.* Umweltbundesamt (Herausg.), Berlin.

USA (1976). The Toxic Substances Control Act (TSCA) PL 94-469.

Weber, A. & Barben, H. (1984). Beurteilung ökologisch relevanter Daten neuer organischer Chemikalien. *Chimia,* **38,** 443.

Weiß, M., Kördel, W., Kuhnen-Clausen, D., Lange, A. W. & Klein, W. (1988). Priority setting of existing chemicals. *Chemosphere,* **17,** 1419–43.

5

Trends in Environmental Hazard and Risk Assessment of Chemicals

FINN BRO-RASMUSSEN

Laboratory of Environmental Science and Ecology, The Technical University of Denmark, DK-2800 Lyngby, Denmark

INTRODUCTION

Hazard is defined here as the potential for a chemical to cause harm or adverse effects. It can normally be described as a function of two elements: the exposure from the chemical and the potential effects resulting from such exposure (OECD, 1986).

An environmental hazard evaluation is often performed in terms of a step-wise procedure (Landner *et al.*, 1982; Bro-Rasmussen, 1987) in which we may distinguish between the initial, relatively simple hazard identification (HI), and the following hazard assessment (HA). Basically, the assessment constitutes a comparison between estimated exposure concentrations of the chemical in the environment and the concentration or dosage levels which may create effects on specified target systems, if they are exposed.

The assessment concludes in identifying targets potentially at risk and relates these to the possible exposure and exposure routes (Fig. 1).

'Risk' is a word which is often confused with the term 'hazard'. It is integrated here in its statistical context, namely as the probability for a chemical to cause unacceptable harm or effects as a result of an exposure. In the assessment of environmental risks, reference is made to specified releases of chemicals, e.g. from production, from use, storage, transport or similar situations. The possible exposure routes and the exposure measured in intensities or concentrations, frequencies and duration are identified and evaluated against the

1. HAZARD IDENTIFICATION (HI)
To identify hazardous properties inherent to the chemicals, e.g.
persistence,
bioaccumulative potential,
toxicity, and
possibly selected properties which indicate environmental mobility and reactivity.
HI concludes in classification or categorization of chemicals in numerical or otherwise prioritized order.

2. HAZARD ASSESSMENT (HA)
To assess the potential for a chemical to cause harm (adverse effects) to targets or target systems, such as human population, environmental species and/or ecosystems. The assessment depends on information to predict
environmental exposure (distribution or pathways fate).
assessment of effects with reference to identified receiving environments (targets at risk).
HA concludes in identification of targets at risk related to potential exposure and exposure routes.

3. RISK ESTIMATION (RE)
To estimate the probability that a chemical causes harm (adverse effects) as a result of a specified production, use and/or other emission into the environment. The estimation involves a comparative study of
environmental exposure (intensity, frequency and duration) and
exposure routes in the study area against data on acceptable/unacceptable effects in exposed target.
RE concludes in a relationship (probabilistic or quality-related) between potential target-effect data and data on acceptable or unacceptable effects.

FIG. 1. The logics of sequential hazard-risk assessments of chemicals.

possibility for damage or unacceptable effects in individual targets or target systems.

The risk assessment concludes in a relationship between potential target-effects data on one side and data on acceptable or unacceptable effects on the other (Fig. 1).

The concept of acceptability has obviously become a crucial element to be considered as a constituent in the evaluation process. It is to be remembered, therefore, that this concept cannot be defined solely on the basis of scientific considerations and advice. It has to include the practices of other professional disciplines, social sciences, public perceptions and, often, political interpretations. The definition of criteria for acceptability of effects in the environment is extremely complex, and it is difficult to establish operational concepts unless we can select indicator- or key-species, as well as ecological key-functions for that purpose (Stephen, 1986).

Provided such a basis is established, we must attempt to distinguish between acceptable effects and unacceptable effects in order to verbalize/visualize the hazards which we impose on the environment, and in order to measure the risks which we may have to take on behalf of the environment.

As humans we are often more willing to accept effects on other species than on man. This is noted, for example, when effects and changes in environmental systems are reported without sanctions or further protective reactions. In the environmental contexts, the risk estimation may in such cases be said to deteriorate into an identification of 'risk qualities', i.e. of types and intensities of effects (which are accepted), rather than being the quantified determination of frequencies and durations of unacceptable exposure, which should otherwise be the main concern and characteristic of risk assessments.

ASSESSMENT AND ACCEPTANCE OF MARINE POLLUTION

The pollution of many marine areas has developed rapidly in recent years into a stage of severe deterioration of coastal waters, and has created many definite and unacceptable changes of the quality of the open sea. This has become an environmental issue of great urgency and great complexity. It has created a situation which increasingly calls for detailed assessments of many specific scenarios. These may vary between local estuaries and fjord systems, and regional seas such as the Mediterranean or the Baltic Sea, and even highly exposed sections of the ocean, for example, the North Sea.

For an analysis of this situation we are confronted with immense difficulties, and we are often hampered in our attempts to identify the complex cause and effect relationships between specific chemical pollution and observed disturbances in the natural system. This is most often the result of lack of sound, scientific knowledge. Problems, however, may in some cases be amplified or widened as the result of professional secrecy, or by reserved attitudes for economic or other similar reasons (even lack of will). It is normally difficult, therefore, to predict individual changes and adverse effects in the marine systems from specific pollution. It may be even more difficult to forecast accurately the results to be expected from pollution control if (or when) regulations are installed and enforced.

In spite of this, there exists today a reasonable consensus of opinion among marine and other experts in their identification of the most important elements of the chain of events, and about the major cause and effect relationships, which describe the increasing number of marine pollution situations.

RISK ASSESSMENT OF SPECIFIC CHEMICALS

A number of directly observable effects, such as fish kills, disappearance of coastal marine vegetation, or increase in number of fish deformities has often been directly linked to the impact of individual chemicals or heavy metals, which are candidates for a general classification as hazardous to the environment (Cairns, 1983). Such cases result from emissions or they are caused by effluents from point sources, e.g. industries, urban sewage outlets or spills from marine vessels. They may result from single (accidental?) discharges, or from regular or intermittent emissions.

In any such observed cases an identification of hazardous properties of individually found chemicals will normally be initiated, e.g. for toxicity, persistence, bioaccumulative potential, etc. Similarly, the hazards of chemical mixtures of effluents have to be initially characterized by their ecotoxicological profiles (Nordforsk, 1985).

A detailed and more comprehensive hazard assessment may follow as a next step in the evaluation procedure. Potential effects are predicted or they are experimentally identified and evaluated in possible target organisms or systems. Potential concentrations are estimated for comparison under differing exposure scenarios, including judgements on worst-case situations based on prediction, on experiments or from monitoring activities.

If the hazard assessment shows that potential concentrations in certain recipients or in specific target systems will exceed critical levels for (adverse) effects, this is an indication that a risk evaluation should be initiated. Thus, the risk assessment for the specific situation is made on the basis of a preceding hazard assessment of the chemical(s), and it represents an advanced stage of the full evaluation process.

ACCEPTABLE/UNACCEPTABLE EFFECTS OF SPECIFIC CHEMICALS

In order to obtain results of a reasonable quality and credibility from the risk assessments, these are—or they should be—based on experience and established evidence, including information from relatively detailed exposure analyses and effects assessments (see Fig. 2).

Methods and technical means for performing valid risk assessments are—at least theoretically—often available for many individual priority chemicals in the environment. Reference is required to production and marketing information as the basis for release matrices and mass balance studies, to analytical techniques and computerized modelling facilities etc. for exposure analyses, and to laboratory testings, monitoring and surveillance data for effects assessment.

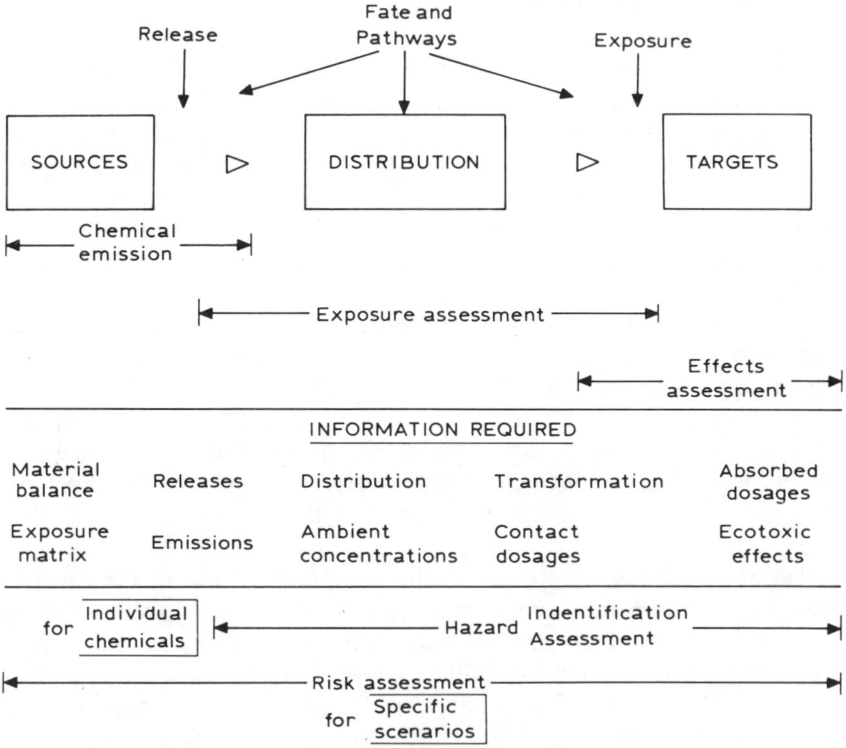

FIG. 2. Elements and conceptual frames to consider in hazard-risk assessments.

It is hard to point to any specific part of the combined procedure as being more (or less) critical as far as the information requirements are concerned, although it may be tempting to mention the lack of chronic ecotoxicity studies for a great number of industrial chemicals. Also disturbing is the continuous uncertainty, which is mostly attached to prediction of long-term effects at the community level or in ecosystems caused by individual chemicals. Figure 3 shows some examples of established and suspected cause and effect chains.

When risk evaluations are brought to their concluding steps, any critical exposure-effect data are related to corresponding levels of acceptable (or unacceptable) effects. This brings the interpretation of acceptability into focus, probably as the most critical part of the total evaluation process. An appropriate distinction between acceptable and unacceptable ecotoxic effects and changes in the ecosystem is critical for the optimum protection of the sea. But many such distinctions are debated scientifically and mostly also in the light of conflicting interests among various groups of society.

For the time being, there seems to be a tendency in many countries to tighten up restrictions on emissions to the marine environment from point sources. It is, however, still the rule to accept a certain level of (adverse) effects in the environment. This can be noticed, for example, in the practice of defining (and accepting) mixing zones around industrial and urban discharges instead of applying zero discharges or best practical means as the environmentally more appropriate strategies.

This observation is clearly connected to the introduction of so-called Modified Quality Objectives as a mode of operation in recipient quality planning in various countries (Miljøstyrelsen, 1983). This means that certain (adverse) effects can be defined as legally acceptable on special terms or as low standard quality. Any such acceptable effects will have to be defined specifically, referring to the type of effects and to the amount of effects (or the extent) which can be accepted.

The uncertainties connected to the acceptability and to the acceptance of environmental effects may also be recognized in the practice of establishment of Quality Objectives for individual chemicals, e.g. List 1 for priority chemicals within national regulations, EEC Directives etc. In the EEC Directive on Dangerous Substances in the Aquatic Environment (EEC, 1976), it is mentioned that a No Effect Level (NOEL) should be taken into account in the setting of

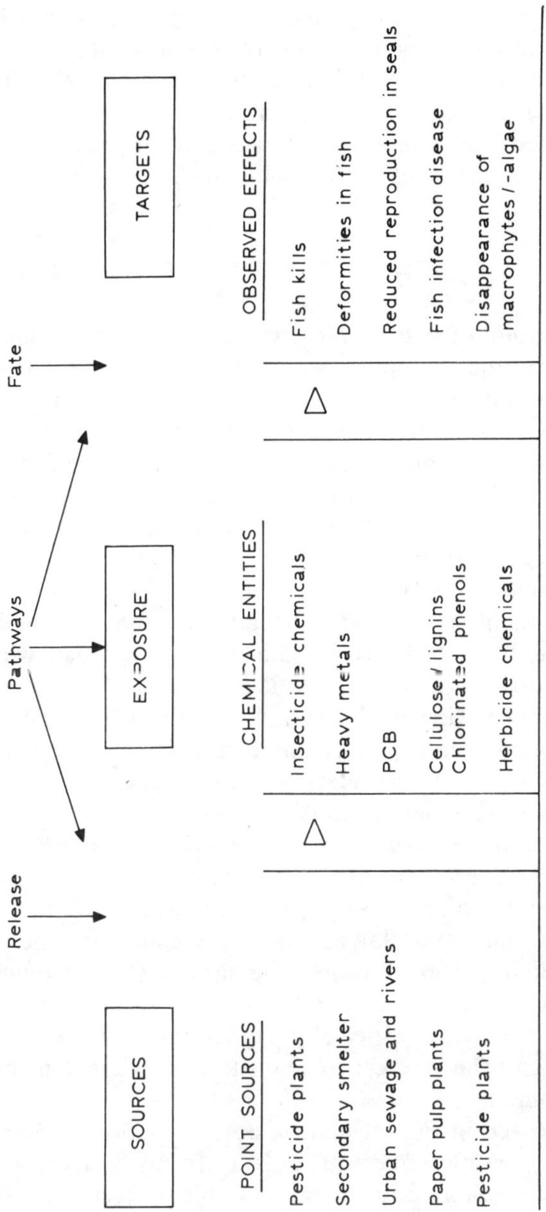

FIG. 3. Some examples of established and suspected cause-effects chains connected to marine pollution.

objectives. This observation, however, is not supported by specifications or guidance on target organisms or biological systems to which it should be applied, nor is it prescribed how the NOEL should be determined. The acceptable (or unacceptable) effect levels are therefore left open for interpretation instead of being defined as an environmentally safe tool in the hands of the risk assessor.

RISK ASSESSMENT OF MARINE EUTROPHICATION

In the description of natural processes and normal biological functions in the sea, primary importance is to be attached to the role of non-toxic, organic (biological) material, of major nutrients such as nitrogen and phosphorus, and of the available supply of oxygen in the water masses. As normal constituents (among numerous others) these substances are essential for the proper structure and functions of the marine system, and as nutrients or sources of energy they can hardly be categorized as intrinsically hazardous in any classification scheme for environmental chemicals.

Nonetheless, it is well-established that uncontrolled inputs of nutrients and biomaterial constitute a pollution load which imposes specific hazards to marine life. This is basically due to excessive primary eutrophication in the surface waters, which creates sedimentation of organic material and an increased oxygen demand—often developing into oxygen depletion—in the bottom zones as the result of the accelerated biological degradation processes.

This is a complex chain of events which has been dramatically demonstrated in recent years in an increasing number of observations of fish kills and avoidance reactions (disappearance) of fish and other species in European coastal waters and enclosed marine areas due to oxygen depletion and bottom inversions. The correlation of this development to the increased rate of nutrient inflow to the sea from land-based activities (especially agriculture) is reasonably well-established, and the interest in realistic risk assessment procedures is correspondingly high.

As summarized in Fig. 4, marine eutrophication is characterized by a number of possible effects. They vary from the increase in primary productivity of (planktonic) biomass, which affects, for example, the competition for specific food resources, to reductions in certain habitats (cf. reduction in coastal vegetation) followed by changes in

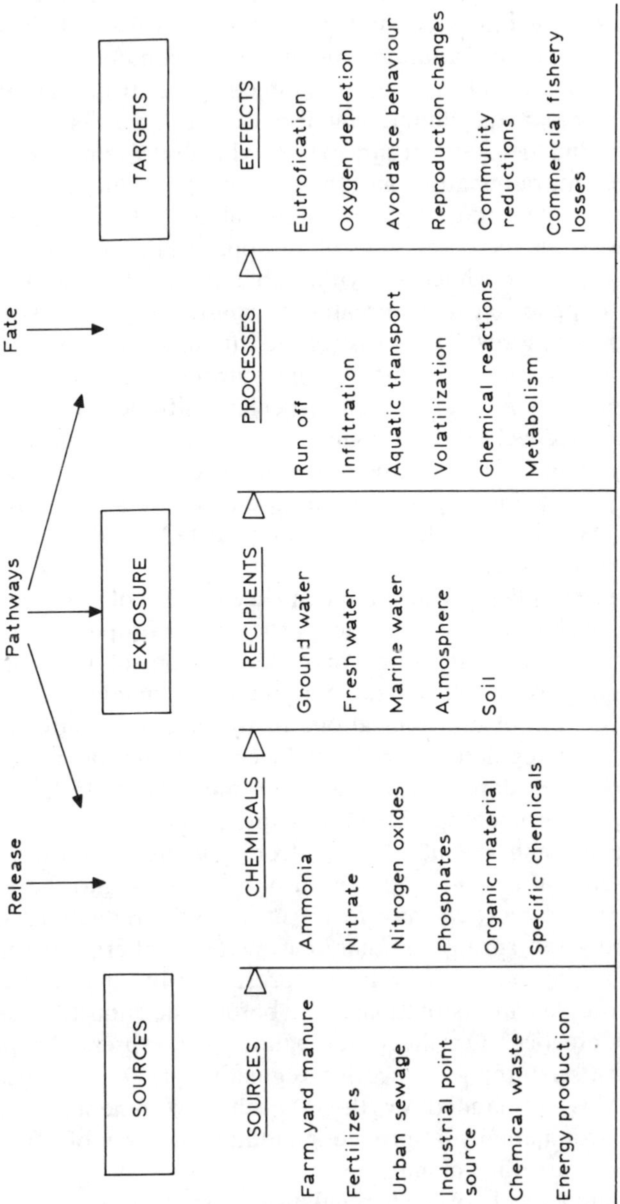

FIG. 4. Important elements to consider in assessment of marine eutrofication.

fish populations. Significant elements of this picture are also the disruptions of the biological and physico–chemical conditions at the bottom due to sedimentation and increased oxygen demand.

Especially important among the effects of eutrophication is, of course, the impact on commercial fisheries. It involves on one side increased fishing yields of certain species which may temporarily thrive from the rich nutrient supply and the increasing productivity in pelagic zones of the open sea, e.g. cod and salmon. This is contrasted, however, with an increased rate of disappearance of several coastal and benthic species which are vulnerable to the fluctuating oxygen supply or depend on the threatened coastal vegetations in their reproduction and growth patterns (cf. eel, flounder and lobster).

For a risk assessment of the marine eutrophication caused by the excessive nutrients flow to the sea, especially nitrogen, it is important to consider selectively the ecological, the commercial, and the public significance of any of the above-mentioned effects. For each of them it is necessary to define the unacceptable effect levels in terms of, for example, survival of single species, undistorted productivity, maintenance of fish habitats, etc.

At the present stage of development in marine pollution control, it is well-known that several countries, at least in Europe, are gradually initiating regulatory restrictions in order to control the outflow of nitrogen and phosphorus to the marine environment, and thereby reduce the degree of eutrophication. In this situation, however, it is noteworthy that regulations are introduced and investments in pollution controls are done mostly without any clarified definition of acceptable and/or unacceptable effects and effect levels.

As a specific and recent example from Swedish–Danish waters it may be noted that a 50% reduction in nitrogen flow into the Kattegat section of the Baltic Sea may result in 15–30% reductions of algal blooms during the spring–summer seasons (Rosenberg, 1986). These figures are interesting because they represent only a partial reduction towards earlier normal situations, i.e. before the industrialization of agricultural practice. On the other hand, they express the practical targets and a scientific guidance for regulation of Danish agricultural practices. This is planned by the Danish government in order to achieve environmentally improved handling and use of commercial fertilizers and farmyard manure.

An estimated result of such regulation is that the oxygen demand may be reduced by a corresponding 30%. The (deficient) average oxygen content in bottom zones of the Kattegat is expected to increase from a critical level of 2·5 mg litre^{-1} to a less critical—although still not normal—level of 2·8 mg litre^{-1}. In the context of risk assessment, the depletion of oxygen in marine bottom zones has thus been defined as an important key effect, and in the specific situation the value of 2·8 mg litre^{-1} becomes significant as a level corresponding to acceptable effects.

CONCLUDING REMARKS

For the validation of the single example which is illustrated here, it remains to be seen whether oxygen depletion is an appropriate key effect for the risk assessment in comparison with other possible effects, e.g. on single species or ecosystem functions.

Equally important, of course, is the question whether the accepted level will meet the expectations (or requirements) of not only commercial fisheries' interest, but also serve the purpose of protection of the overall quality of the marine environment.

Obviously, such questions are raised as universally important in any risk assessment. However, in this specific case the questions also express an informed doubt that the 50% reduction of nitrogen outflow into the sea may not be sufficient to ensure a restoration of the marine ecological quality. It may merely reduce the rate of a continued man-made eutrofication process.

If this judgement proves correct, we may, in fact, have failed to eliminate the risks of adverse effects caused by eutrofication and to have only restricted a process by accepting effects and effects levels which should have been deemed unacceptable. It may be said that we have changed the type and extent of potential hazard and we may have altered the rates of development, but without reducing statistically the 1:1 (or 100%) risk of damage to the marine ecosystem.

REFERENCES

Bro-Rasmussen, Finn (1987). Application of protocols for hazard assessments. In *Attitudes to Toxicology in the European Community*. ed. P. L. Chambers, John Wiley & Sons, Chichester.

Cairns, John, jr. (ed.) (1983). *Ecoaccidents. NATO Conference Series. Series 1: Ecology*, Vol. 11. Plenum Press, New York.

EEC (1976) Directive 76/464, 4 May 1976, on Dangerous Substances for the Aquatic Environment, referring to Environmental Protection Terminology in Official Journal of the European Communities No. c 112/49 of 20 December 1973.

Landner, Lars, et al. (1982). Systems for testing and hazard evaluation of chemicals in the aquatic environment. ESTHER-report no. snv pm 1631, 1 December 1982, National Swedish Environment Protection Board.

Miljøstyrelsen (1983). Guidelines for recipient quality planning (in the Danish language). Part 2, Coastal waters. National Danish Environment Protection Agency, DK-1402, Copenhagen, Denmark.

NORDFORSK (1985). Ecotoxicological methods for the aquatic environment, Subproject no. 4, Characterisation of industrial waste water. Nordforsk, Helsinki.

OECD (1986). Report of the OECD Workshop on Practical Approaches to the Assessment of Environmental Exposure, 14–18 April 1986, Vienna, Austria.

Rosenberg, R. (ed.) (1986). The situation of eutrofication in Kattegat (written in the Swedish language). Report no. 3272, November, National Swedish Environment Protection Board.

Stephan, C. E. (1986). *Proposed Goal of Applied Aquatic Toxicology, Aquatic Toxicology and Environmental Fate,* Vol. 9, eds T. M. Poston & R. Purdy. ASTM STP 921. American Society for Testing and Materials, Philadelphia, pp. 3–10.

6

Managing the Environment: Role and Duties of the Chemical Industry in an International Context

L. JOURDAN
CEFIC, 250 Avenue Louise, 1050 Bruxelles, Belgium

Conseil Européen des Fédérations de l'Industrie Chimique (CEFIC) is the European Chemical Industry Federation representing the chemical industry in 15 countries in Western Europe, an industry which employs two million people and accounts for 30% of the world production of chemicals.

CEFIC has a mandate from its members, the national chemical industry associations, to deal with all international issues, whether at EEC level or concerning the Economic Free Trade Association, the OECD or the UN organizations. This embraces a wide range of matters, such as international trade, industrial property, raw materials and energy supply. In this range of activities, issues related to the impact of the chemical industry on health, safety and the environment obviously have a high ranking on the list of priorities. CEFIC, therefore, is well-placed to identify the new trends in environmental management and, in particular, in 'clean technologies', the topic of the present session.

Like any trade organization, CEFIC has a dual role. First, it represents the interests of its members, the chemical companies, through their national associations *vis-à-vis* regulations prepared by the international authorities, e.g. European Community Directives, OECD Recommendations. Conventions of the UN agencies, etc. As these regulations are often seen as a constraint to free enterprise, this activity may be regarded as being of a 'reactive' nature. However, it is now more and more accepted that a comprehensive and reasonable set of international regulations is necessary to protect the environment, and CEFIC contributes to the establishment of these regulations in making available its expertise to the authorities, whilst taking care that these regulations are justified and compatible with sustainable development, as outlined in the Brundtland Report.

Second, in order to enhance its credibility with authorities as well as to meet the spirit of 'self-regulation' deeply entrenched in the entrepreneur's mind, CEFIC assists chemical companies in developing their policy in health, safety and environment matters. This takes the form of Principles for Environment Protection, Guidelines and Codes of Good Management, for example in the field of waste management, safety in warehousing, etc., or Programmes for Risk Communication, and many others. Through this 'pro-active' role, CEFIC contributes to bridging the gap that has developed between the public perception and industry's actual performance, this gap being responsible for the broadly negative image that the chemical industry has to face at the present time.

Before speaking about trends in environmental management, let us try to identify some features of the chemical industry which make it somewhat unique compared to other industries.

(1) *The products of the chemical industry are not directly visible to the public.* With very few exceptions, chemicals produced by the chemical industry are not directly sold to the public, but to processors who transform them into products of day-to-day life. The public therefore does not see the link between the gases, liquids or powders produced by the chemical industry, which are often responsible for well-publicized accidents and environmental damage, and objects which help to make life better. Who in the public could associate a nice piece of ladies' underwear with caprolactam and hydrogen cyanide? A glossy car paint with isocyanates and phosgene? A toothpaste with acid wastes from titanium dioxide manufacture, etc.?

Maybe the chemical industry itself is responsible for this situation because of its long-standing insufficient communication, especially with the public. But certainly this lack of visibility of the chemical industry's products has greatly contributed to the gap mentioned earlier.

(2) *None of the chemical industry's processes or products is totally environmentally innocent.* Of course, this holds true for most human and industrial activities, which all have a certain unavoidable impact on the environment. But in addition, the

chemical industry uses processes where high levels of energy, temperature or pressure are involved, with the potential risk of deviation from controlled situations, and therefore of accidents. It carries out chemical reactions, i.e. transformation of molecules, leading, besides the desired ones, to others which are of no or low value and which may be difficult to handle. Finally, many of the intermediates used or the products manufactured are hazardous because of their physical properties (flammability, explosivity), their biological activity (corrosivity, toxicity) or their impact on the environment (ecotoxicity, persistence, bioaccumulation). It should be recognized that more often than not these hazards are an intrinsic feature of the product itself: no bleaching agent could have bleaching effects without being an oxidizer, no pesticides would control pests without being toxic to the considered target. When chemicals are properly handled, their hazards have no or little impact on man and the environment. But there may be accidental releases during manufacture, storage or transport (e.g. the catastrophic accidents at Bhopal, Basel or Los Alfaques); chemicals may be wrongly used (e.g. the eutrophization of lakes and rivers because of excessive use of fertilizers); finally at the end of their life or that of products containing them, chemicals may become an awkward waste (e.g. PCB in transformers, CFC in refrigeration units).

(3) *The chemical industry is deeply involved in international activities.* Some of its raw materials are imported from remote locations. Most of its finished products, because of their high added value, can be exported to the remotest countries. Most companies have worldwide marketing strategies, even small and medium-size ones, whilst all large companies are 'multinationals' having production facilities and sales offices in many countries in the various continents.

(4) *The chemical industry has always been strongly innovative.* Every day new molecules are invented to meet new requirements from society or to improve the performance or safety of specific uses. New processes are developed to manufacture products cheaper, safer and with less impact on the environment. R and D expenditure of chemical companies constitutes a significant percentage of their total turnover, whilst the relatively short life of some products and processes requires high capital investment costs.

Within this picture the responsibilities of the chemical industry for the protection of the environment are easy to identify.

A first step is 'risk analysis', i.e. the various methods to evaluate the hazards presented by a product or a process, to consider the risks they may pose to the environment, and to reduce these risks to as low as possible a level, considered as acceptable.

For chemicals, this means testing their physical, toxicological and ecotoxicological properties before placing them on the market. Because of its importance to the protection of man and the environment, and to avoid distortions of competition in view of the high costs involved, this is an area strictly governed by national or international regulations (the EC Sixth Amendment, US TSCA, etc.).

Beyond this evaluation of inherent hazards, risks posed by chemicals to man and the environment are assessed considering their volume of production, their immediate or delayed effects, their fate in the various compartments of the environment during their life-cycle including at the final stage, and their economic worth and value to society. On the basis of this assessment the decision may be taken, either voluntarily by the producer, or through governmental or intergovernmental regulations, to prohibit or restrict their marketing. See for example the Montreal Protocol restricting the use of CFCs.

As fas as processes are concerned, a number of methods are currently in use to identify hazards from the various steps of the process or items of equipment, to evaluate the probability of failure or deviation from normal conditions, and to estimate the consequences of such events. But also the impact on the environment under normal conditions (emissions, waste, noise, etc.) is assessed and steps are taken, at the research, development and construction stages, to minimize these risks and impact down to an acceptable level. If this is not technically or economically feasible, the decision should be taken not to use the process. This is mostly a decision resting with the industrialist, even if a number of regulations now set the objectives and modalities of such an assessment (e.g. the EC 'Seveso' Directive).

Another responsibility is 'risk management'. When a process is operated or when a chemical is marketed, the risks they present to man and the environment have to be kept at the accepted level.

For processes, this means that technical and managerial measures are taken to keep the process and equipment under strict control, that any deviation is immediately detected and corrected, and finally that emergency procedures are established to limit the consequences of accidental events.

Managing the risk of chemicals relies on clear and precise information provided to users, through labels, material safety data sheets, education and training of processors in the safe and correct use of chemicals, including waste disposal. Incidentally, producers and processors might jointly be involved in the establishment of waste recovery or disposal schemes, such as spent catalysts or CFCs from used refrigeration units.

Exporting a chemical or transferring a technology to a remote country, especially a developing one, can on no account be an excuse for not fully implementing these responsibilities. Manufacturers and traders exporting chemicals should ensure that relevant information on the safe use, handling and disposal of their chemicals reaches their customers in importing countries; this information should be in a format appropriate (e.g. symbols and languages) for the importing country. They also should undertake or support training and educational activities in importing countries designed to improve customers' and users' understanding of the information provided.

When transferring a technology the supplier should give all information allowing the client to achieve a level of safety and protection of the environment equivalent to the installation of the supplier. This does not mean that the same technical measures have to be taken; they have to be adapted to the local conditions (culture, climate, geology) of the receiving country.

But risk management is not a steady state: public demands or new findings might render unacceptable a level of risk which had previously been considered acceptable. This is not new, it has been a characteristic of the chemical industry since early industrial development. The Leblanc process for the production of soda ash was a major breakthrough at the end of the 18th century, because it opened the door to industrial production of alkali under much better conditions than from the washing of wood ash. But the large quantities of hydrochloric acid discharged by the process made it no longer acceptable, economically as well as environmentally, when the Solvay process was developed some 100 years later.

White lead had been extensively used in the 19th century because of its excellent weather-resistant properties in paints for buildings. But when its toxicity was discovered and confirmed, its use was completely discontinued at the beginning of the 20th century—admittedly not without some social problems. In the meantime an even better white pigment had been developed, titanium dioxide.

This is where the innovative capacity of the chemical industry comes in. Maybe more than other industries, the chemical industry has always been prepared to respond to the challenges set by society. Most of these challenges have been of an economic nature, the public calling for better products at lower prices. More and more now they are driven by environmental considerations: the research programmes to find substitutes for CFCs are a typical example. It also happens that environmental and economic considerations both contribute to the development of a new chemical or process. This was the case with the Solvay process for alkali production and also held true for most of the new or improved processes which were so designed as to reduce their emissions, by-products or waste, the so-called 'low-waste technologies'.

There are several ways in which a low-waste or clean technology can be developed:

(1) Operating conditions may be tuned up to optimize the global yield of the reaction, reducing the generation of unreacted or intermediate products.
(2) Unreacted or intermediate products may be recycled in the process, used as raw materials or intermediates in another step of the process or in another process, or recovered and shipped to another client.
(3) A totally different process may be developed, generating by-products in lower quantity or that are easier to re-use or recover.
(4) The remaining emissions may be treated to lessen their environmental impact.

To reach a specified objective the choice from amongst these various possibilities will obviously be governed by economic factors. Nobody will dispute that the most successful and profitable companies are those whose innovative capacity allows them to develop highly efficient processes, or complex integrated plants where several units are linked with each other by a delicate balance of residual raw materials, or new routes for manufacturing a chemical, or finally a chemical of similar benefit to society but with less impact on the environment during its manufacture or use.

But let us be realistic. Such innovations cannot be developed overnight, most of them are the result of long and tedious research. It should also be realized that a too rapid turnover in products or processes would simply mean that the capital investments incurred would not have the time to be written-off, thus dramatically increasing the cost of the substitute product or process. And above all, one should never forget that however sophisticated and efficient a technology may be, there will always remain a residue of a certain nature: this is a physical law, apparently ignored by those spreading among the public the idea that industry could operate without waste.

And this introduces the final responsibility of industry developed in this paper, its communication with the public. For too long, industry and the chemical industry in particular maintained an attitude of secrecy, possibly based on the assumption that the public was not able to understand the nature of its operations. The same reluctance towards dialogue and transparency existed with authorities and the labour force.

This time now is over. Most companies have adopted a policy of openness with public authorities, have developed programmes of internal communication within their personnel, and, above all, have begun discussing with the public, either directly, through open doors, media news, press conferences, or through its representatives, like local authorities, community leaders and pressure groups. Of course, not everything can be freely disclosed, there is some information that has to be kept secret in order to protect competitive positions or public security. Defining what effectively constitutes confidential information is a difficult exercise calling for skill and lucidity.

In the field of health, safety and the environment the matter is controlled by a number of regulations. One of them deals with the information to be given to the local community possibly affected by an accident from a chemical plant; here the public need to be told how they must behave in case of an emergency; they would do so even better if told about the nature of the industrial operation and if they have confidence in the safety measures taken by the industrialist. To meet the regulatory requirements or to go beyond them with a view to developing good relationships with the community and gaining a better image, many companies or industrial sites are presently carrying out such programmes of risk communication. This is only one aspect of the overall issue of communicating with the public on environmental matters, maybe not the easiest, but generally speaking, experience so far has been broadly positive.

To sum up, risk analysis and risk management of both products and processes; innovation for developing low-waste technologies and substitutes; communication with the public on environmental matters:

these are the most significative trends in the responsibilities of the chemical industry when managing the environment. You will not be surprised, of course, to find them amongst the 'CEFIC Guidelines for the Protection of the Environment' referred to at the beginning of this chapter.

7

Recent Developments in Pollution Control Technology for Boilers, Furnaces and Incinerators

JUMPEI ANDO

Faculty of Science and Engineering, Chuo University, Kasuga, Bunkyo-ku, Tokyo 112, Japan

FLUE GAS DESULFURIZATION (FGD)

Overview

Global SO_2 emission by human activity has now reached about 100 million tons yearly. Flue gas desulfurization (FGD) has become popular in many countries, and with a total capacity of 170 000 MWe, is expected to remove nearly 10 million tons of SO_2 in 1989. There are currently about 2000 FGD plants in existence for boilers, furnaces, etc. that remove more than 80% of the SO_2 produced. In addition, lime is injected at about 1000 municipal solid waste (MSW) incinerators to remove 40–90% of acid gas. Over 90% of FGD plants are installed in three countries: the USA, Japan and the FRG (Table 1). The process is also used in Austria, the Netherlands and Sweden.

Lime/limestone processes are used for more than 90% of the capacity because limestone is the cheapest and most abundant absorbent of SO_2. Lime/limestone processes will continue to be the major FGD process.

Major Wet Processes

Wet processes are applied to more than 90% of the total FGD capacity because of the high SO_2 removal efficiency. Large boilers and furnaces use mainly wet limestone scrubbing to remove 90–95% of SO_2. The process is simple and is widely used in the USA. However, it is not used in other countries because of the requirement of a large landspace for disposal of the sludge produced.

TABLE 1
FGD Capacities in Three Countries (1000 MWe, 1989)

	USA	*Japan*	*FRG*
Wet lime/limestone process			
Waste sludge production	50	0·2	0
Gypsum production	6	32	32
Other wet processes			
Waste by-product	Uncertain	9[a]	0
Useful by-products	3	9[b]	2
Semi-dry and dry processes	6	6[c]	4
Lime/limestone process	0	0·3	0·2
Other processes	65	57	38

[a] Mainly sodium and magnesium sulfate liquor.
[b] Mainly sodium sulfite for paper mills.
[c] Mainly for municipal solid waste incinerators.

Lime/limestone scrubbing to by-produce saleable gypsum was developed in Japan and has also been used in the FRG, Austria, the Netherlands, China, etc. The process removes 90–98% of SO_2 using 0·92–1·0 mol CaO to 1 mol SO_2 and by-produces gypsum, which is useful for wallboard, cement, plaster, etc.

Japan utilizes nearly all of the FGD gypsum, about 2 million tons yearly, while the total annual gypsum consumption is 7 million tons. The utilization of FGD gypsum has also become popular in the FRG and has started to catch on in the USA, etc. (Maknsi & Ellison, 1988). Even if it needs to be thrown away, gypsum may be a better product than the sludge.

Wet sodium and magnesium scrubbing is used at more than 1000 small plants, mainly in Japan, to by-produce saleable sodium sulfite or waste sodium and magnesium sulfate liquors that are released into the sea (Ando & Kaplan, 1988).

Problems with Wet Processes and Counter-Measures

Wet processes have two common problems: they require gas reheating and they produce wastewater.

The flue gas usually leaves a wet scrubber at 50–60°C and is reheated to 70–120°C before emission from a stack. To save fuel used in gas reheating, gas–gas heaters are used mainly for FGD plants for coal-fired boilers to raise the temperature from 50°C to 80–110°C using the energy of the flue gas at 130–200°C.

The wastewater is usually neutralized to precipitate out and separate

TABLE 2
Typical Examples of SO_2 and SO_3 in Gases

Fuel		*Flue gas at ESP outlet*			*Wet FGD outlet*		*Semi-dry FGD outlet*[a]	
Type	*S (%)*	*O_2 (%)*	*SO_2 (ppm)*	*SO_3 (ppm)*	*SO_2 (ppm)*	*SO_3 (ppm)*	*SO_2 (ppm)*	*SO_3 (ppm)*
Coal	3	6	2100	10	150	5	300	2
Oil	3	6	1500	50	100	25	200	7
Oil	3	2	1500	25	100	12	200	4
Oil	0·2	2	100	2				

[a] At fabric filter outlet.

heavy metals. In Japan, the wastewater is often subjected also to COD removal. Usually an FGD unit for a 300–500 MWe coal-fired boiler gives 5–20 tons/h of wastewater. Efforts are in progress to eliminate wastewater (Delleney & Dickerman, 1988). One of the ways is to spray the water into the flue gas upstream of an electrostatic precipitator (ESP) to dry it up.

Another problem is the relatively low SO_3 removal efficiency of 40–60% (Table 2). SO_3 levels are low in flue gases from pulverized coal because it is absorbed by fly ash but they are high in the gas from high-sulfur oil. On cooling, SO_3 readily forms fine mists of concentrated sulfuric acid, which are not removed efficiently by conventional wet FGD.

Sulfuric acid mists cause corrosion and plugging of air preheaters and gas–gas heaters, and when SO_3-rich low-temperature gas is emitted from a low stack, the mists may have adverse effects on the adjacent environment.

The formation of SO_3 in the combustion gas is reduced by lowering the excess air (Table 2). Wet electrostatic precipitators (ESPs) are installed downstream of some wet scrubbers in Japan to remove SO_3.

Dry and Semi-dry Processes

Various dry processes have been studied but most of them were not developed commercially because of low efficiencies, complexity, or high cost. Recently in Europe, a semi-dry lime slurry spray drying process has been developed, which removes 70–95% of SO_2 using a fabric filter and 1–2 mol CaO to 1 mol SO_2. The process is also used in the USA for boilers and in Japan for MSW incinerators.

Dry lime/limestone injection is in limited use for boilers because of the low removal efficiency of 30–50% using 1–3 mol CaO to 1 mol SO_2. The process has become popular for MSW incinerators to remove 40–80% of HCl and SO_2. In fluidized bed combustors, limestone is added to remove 50–90% of the SO_2 using 1–4 mol CaO to 1 mol SO_2.

These dry and semi-dry processes require no gas reheating, produce no wastewater, and remove SO_3 at a high efficiency when used with a fabric filter (Table 2). On the other hand, the by-product, a mixture of fly ash, calcium sulfate and sulfite, and lime, is abandoned. Lime may be leached out of the waste. The processes will be used more widely if the by-product can be utilized.

FGD Cost

Wet limestone–gypsum process plants in Japan in 1978–83 cost more than 20 000 yen ($160)/kW and consume more than 2% of the power generated by each boiler. The process has been significantly simplified by the use of a single-absorber, in-situ oxidation process resulting in the reduction of cost to 10 000–15 000 yen ($80–120)/kW and the power consumption to 1–1·5% (Ando & Kaplan, 1988). The cost of removing 1 ton of SO_2 at 90% efficiency by the simplified process, including 15% of investment as an annual fixed cost, may be about $700 for 1000 ppm inlet SO_2 (Fig. 1) and $350 for 3000 ppm.

Compared with the limestone–gypsum process, semi-dry lime spray drying may cost less for 80% removal with low-sulfur coal but may cost more for over 90% removal or with high-sulfur coal. Dry

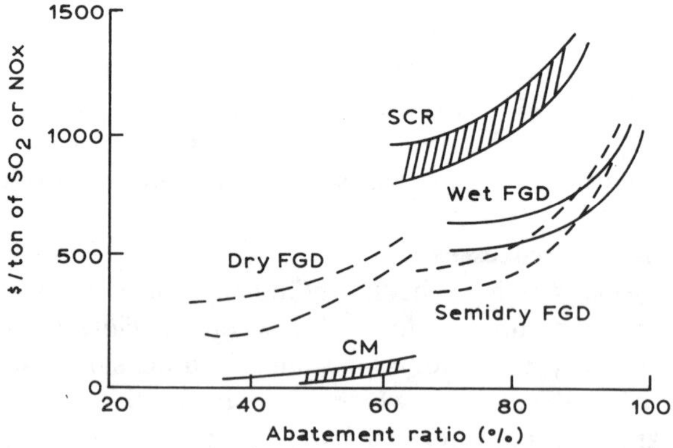

FIG. 1. SO_2 and NO_x abatement efficiency and cost.

lime/limestone processes may cost less to remove less than 50%. The cost varies considerably with the by-product disposal cost and also with the reliability of the process.

Sodium and magnesium scrubbing plants that by-produce waste liquor cost about 5000 yen ($40)/kW in investment but may not be useful in treating large amounts of gases because of the high absorbent costs. Processes that by-produce elemental sulfur and sulfuric acid are costly and have not become popular.

NO_x AND COMBINED SO_2 AND NO_x ABATEMENT PROCESSES

Combustion Modification (CM)

In combustors, combustion modification (CM) can be used to reduce NO_x levels by about 70% for minimum cost (Fig. 1). Various types of CM have been developed, including low excess air combustion, flue gas recirculation, two-stage combustion, and low-NO_x burners. By combining these, NO_x levels may be lowered to 150–250 ppm for coal-, 80–150 ppm for oil-, and 40–80 ppm for gas-fired boilers.

The reduction of excess air is particularly important because it reduces not only NO_x but also SO_3 and fuel consumption, although precise combustion control is required to prevent incomplete combustion.

A remarkable example is the glass-melting furnace of Asahi Glass Co. in Japan. Flue gases from glass-melting furnaces are usually rich in NO_x (1000 ppm or more) because of the long gas retention time at a high temperature (1600°C) and the high O_2 content (around 5–6%) at the furnace (O_2 content at the stack reaches 12–13% due to air leakage through heat regenerators). Asahi Glass has succeeded in lowering NO_x levels to 400 ppm with 20–30% fuel oil saving by changing the burner and by reducing O_2 levels to less than 2% at the furnace (8–9% at the stack) with precise control.

Another example is a cement kiln. A conventional coal-fired rotary kiln with a suspension preheater usually produces a flue gas containing 700–1000 ppm NO_x with 7–10% O_2. Reduction of O_2 to 1·5–2% reduces NO_x by 10–20% and fuel consumption by 5–10% (Table 3). By adding a furnace between the kiln and the preheater to convert calcium carbonate to lime, NO_x can be lowered to 400–600 ppm while the production capacity can be doubled (NSP kiln, Table 3).

TABLE 3
Examples of Combustion Modification for Cement kiln

	Coal (ton/h)		O_2 (%)	Clinker (t/h)	Energy (kcal/kg)	NO_x (ppm)
	Kiln	Furnace				
SP kiln	9	0	8–10	70	900	800–1000
Improved SP	8	0	1·5–2	70	800	650–800
NSP kiln	6·4	9·6	1·5–2	140	800	400–600

For boilers and heating furnaces, 10–20% NO_x reduction and up to 15% fuel saving may be achieved by O_2 reduction with precise excess air control.

NO_x Removal Processes

Selective catalytic reduction (SCR) using ammonia and catalyst at 300–400°C has been developed in Japan and is used at about 250 plants with a total capacity of nearly 40 000 MWe mainly for oil- and coal-fired utility boilers. By using 0·6–0·95 mol NH_3 to 1 mol NO_x, 60–90% of NO_x is removed. SCR is also used in the FRG, Austria and the Netherlands in coal-fired utility boilers and in the USA in industrial boilers.

Dust plugging and SO_3 poisoning, major problems with SCR, have been solved by the development of parallel flow type catalysts based on TiO_2. The catalyst life is 4–5 years for coal- and 7–10 years for oil-fired boilers. The high-quality catalysts usually cost about 3 million yen ($24 000)/$m^3$. HCl-resistant catalysts for MSW incinerators, arsenic-resistant catalysts for wet bottom coal-fired boilers, and zeolite-based catalysts for a gas with low SO_x concentration are also in commercial use.

SCR is simple and reliable but is costly (Fig. 1) and should be used in combination with CM when CM is not sufficient. Another problem with SCR is the disposal of spent catalyst. Valuable heavy metals such as tungsten may be recovered economically. TiO_2 might also be recovered because the content reaches 70–80%, while titanium ores usually contain about 50% TiO_2.

Selective non-catalytic reduction that uses ammonia at 900–1000°C without catalyst has been applied mainly to SO_x lean gas from small boilers and furnaces. Usually 30–50% of NO_x is removed using 1–2 mol NH_3 to 1 mol NO_x, giving more than 10 pppm slip NH_3.

Combined SO_2 and NO_x Removal

The best established way to remove both SO_2 and NO_x from coal-fired boiler flue gases is the application of SCR at the boiler economizer outlet at 330–400°C followed by air preheater (APH), electrostatic precipitator (ESP), gas–gas heater (GGH), and wet FGD (Fig. 2(a)). This system has been applied to many coal-fired boilers in Japan, the FRG, Austria, etc. Low-oxidation catalysts are used and slip NH_3 is usually kept below 3 ppm to prevent the deposit formation in APH and GGH and the contamination of fly ash.

Tail gas treatment as shown in Fig. 2(b) has been used since 1976 for iron-ore sintering machine and sewage treatment sludge incinerators, because the gas temperature ranges from 150 to 250°C and the direct application of SCR causes ammonium bisulfate deposits on the catalyst.

A system similar to that shown in Fig. 2(b) without a wet ESP is now being used for coal-fired boilers in the FRG. The system was used in Japan in 1975–80 for an oil-fired boiler but was given up because of the deposits in the gas–gas heater due to the high SO_3 content of the gas leaving the wet FGD (Table 2). The system may work well for coal-fired boilers because of the low SO_3 content of the gas (Table 2).

The tail gas treatment system is costly because of the fuel requirement for gas heating but may be needed for gases at low temperatures and rich in catalyst poisons or when the SCR reactor cannot be retrofitted at the boiler outlet at 300–400°C.

Simultaneous SO_2 and NO_x Removal

Various simultaneous removal processes have been tested in many countries but most of them have not been commercialized because of

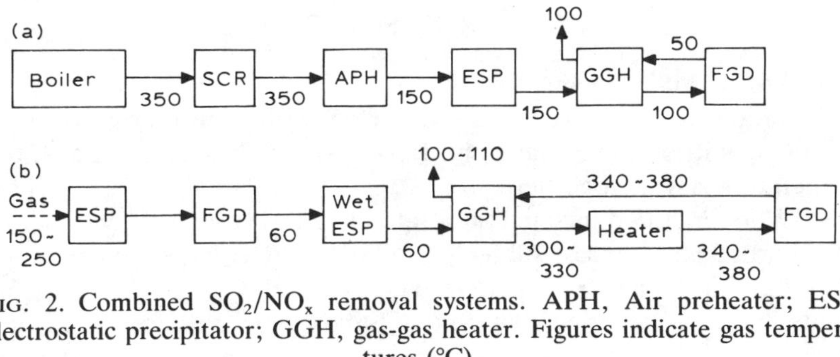

FIG. 2. Combined SO_2/NO_x removal systems. APH, Air preheater; ESP, electrostatic precipitator; GGH, gas-gas heater. Figures indicate gas temperatures (°C).

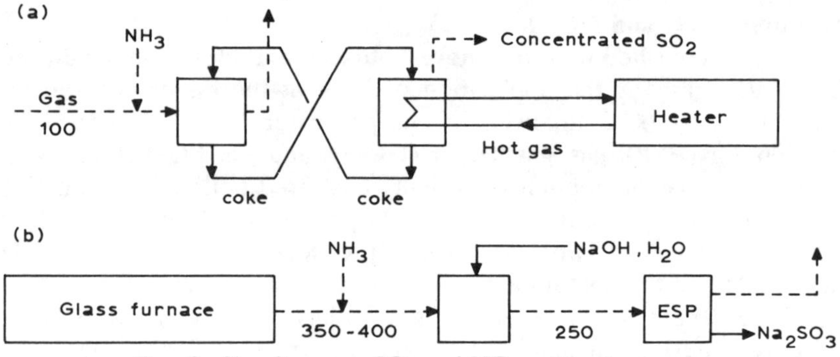

FIG. 3. Simultaneous SO_2 and NO_x removal process.

high costs, low efficiency or problems in operation. Wet processes are applied to about 20 small plants mainly in Japan but not to large plants because of the wastewater treatment problem.

Three commercial activated-coke process plants, two in Japan and one in the FRG, were put into commercial operation in 1986–7. These use ammonia for NO_x removal and by-produce concentrated SO_2 to produce elemental sulfur or sulfuric acid (Fig. 3(a)). The process may not be suitable for SO_2-rich gases because of the high coke consumption.

In a unique process used at a glass-melting furnace in Japan (Fig. 3(b)), ammonia is added to the flue gas at around 400°C and the gas is then sprayed with sodium hydroxide solution. Powdery sodium sulfite containing no nitrogen compounds is caught by an ESP and oxidized by air to sulfate; this is fed to the glass furnace. Ammonia reacts nearly stoichiometrically with NO_x to form N_2.

POLLUTION CONTROL FOR INCINERATORS

Overview of MSW Incinerators

Incineration of municipal solid waste (MSW) has become popular in several countries. The status in Japan, which has more than 2000 incinerators is described here.

MSW is generated in Japan at a rate of 120 000 tons/day, with about 70% burned and the rest subjected to landfill. The largest incineration units have a capacity of 500 ton/day, while the largest incineration plants have six 400-ton/day units. Stoker-type furnaces are most popular while fluidized-bed furnaces and rotary kilns are also used.

TABLE 4
Capital Cost of MSW Incineration Plants (10^6 yen/ton/h)

	Large[a]	*Medium*[a]	*Small*
Dry, semi-dry scrubbing	200–300	300–600	200–500
Wet (sodium scrubbing)		400–700	400–700

[a] With power generation.

Larger incineration plants generate electric power with capacities ranging from 750 to 15 000 kW.

MSW usually produces 6–14 N m^3/kg of flue gas containing 40–100 ppm SO_x, 150–250 ppm NO_x, 300–800 ppm HCl, and 1–3 g/N m^3 of dust with 8–14% O_2. ESPs are mainly used for dust removal, although fabric filters are used at a few plants. The fly ash as well as the bottom ash are usually used as landfill because hazardous materials are in very low concentrations. For safety, the fly ash is pelletized using cement at some of the plants.

Survey works were made on 71 plants with 161 incinerators constructed by 13 major manufacturers. The plant capacities ranged from 70 to 2400 tons/day and the unit capacities ranged from 1·5 to 21 tons/h. Most plants remove 50–80% of acid gas (SO_x and HCl) by dry or semi-dry lime scrubbing, while wet sodium scrubbing is applied at medium-sized plants to remove more than 90% of the acid gas. Table 4 shows the capital cost of the plants, including civil works, pollution control facilities and office buildings. The cost ranges from 200 to 700 million yen ($1 600 000–5 600 000)/ton/h according to the size and type of the plants.

Gas Cleaning Systems for MSW Incinerators

NO_x is usually lowered from 150–250 to 70–150 ppm at 12% O_2 basis by reduction of O_2 from 10–14 to 7–8% and by lowering the combustion temperature from 900–1000 to 750–900°C. Wastewater at the plant is usually sprayed into the furnace to lower the temperature and to reduce wastewater discharge.

At some of the fluidized-bed furnaces, limestone is used as a bed material. At a few large plants, powdery limestone is injected into the furnace at around 900°C (Fig. 4(a)). By those methods, about 50% of both SO_x and acid gas are removed at a lime stoichiometry of about 2.

At most plants, slaked lime in powder or slurry form is sprayed into

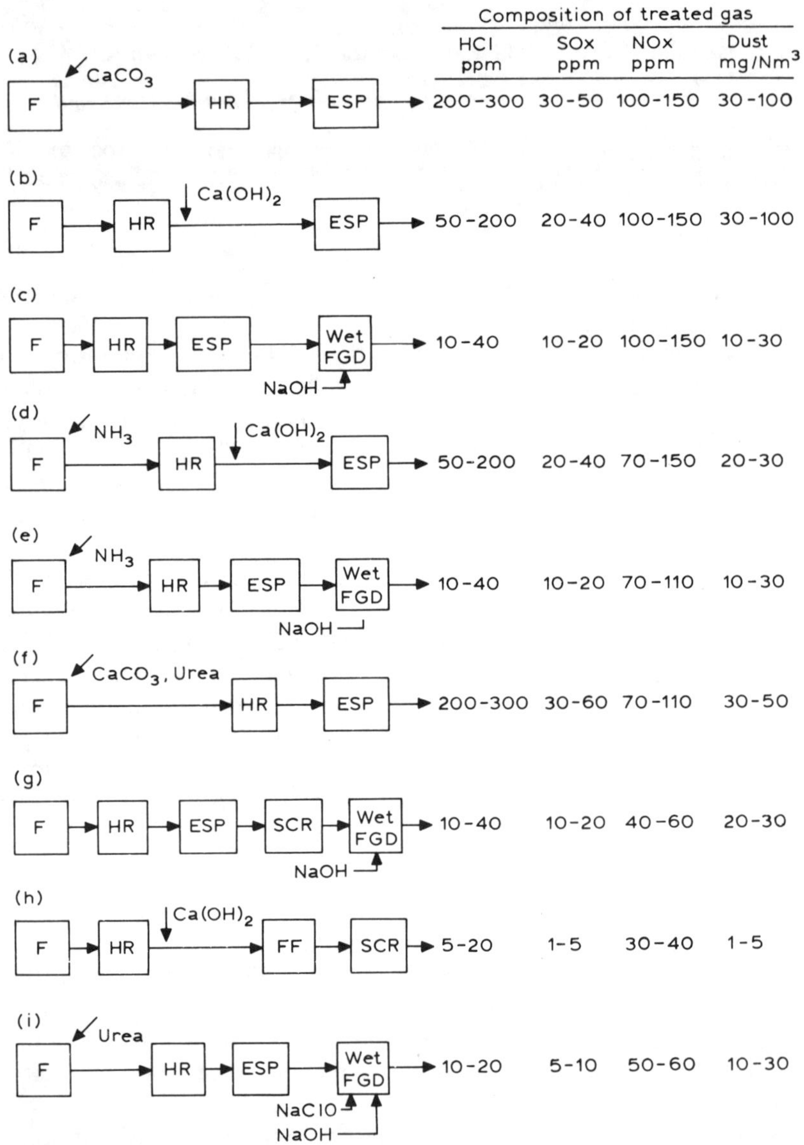

FIG. 4. Gas cleaning systems for MSW incinerators. F, Furnace; HR, heat recovery; FF, fabric filter.

a duct or a reactor at 250–400°C to remove 70–95% of SO_2 and HCl at a stoichiometry of 2–2·5 (Fig. 4(b), (d) and (h). The efficiency is higher with slaked lime slurry than with powdery slaked lime and with a fabric filter than with an ESP.

Wet sodium scrubbing is used at medium-sized plants to remove 90–98% of SO_2 and HCl. At one of the plants a wet ESP is installed downstream of a scrubber to remove even more.

For NO_x removal, selective non-catalytic reduction (SNR) is applied at about 30 plants (Fig. 4(d) and (e) to remove 30–40% of NO_x using 1–1·5 mol NH_3 to 1 mol NO_x. Plume may form by using a larger amount of NH_3. At a few plants, urea solution is sprayed into the furnaces at around 900°C to remove about 30% of NO_x (Fig. 4(f) and (i)) by using 0·5 mol urea to 1 mol NO_x. SCR is used at two plants to remove 50–80% of NO_x at 200–250°C using an HCl-resistant catalyst (Fig. 4(g) and (h)).

Problems with Mercury

Mercury (Hg) is derived mainly from electric cells and is present at levels of 0·03–0·3 mg/N m^3 in MSW combustion gas. Up to 70–80% of Hg is removed by conventional wet sodium scrubbing, while dry and semi-dry processes may remove 50–70% with a fabric filter and less with an ESP. The waste gas leaving a stack usually contains 0·01–0·1 mg/N m^3 of Hg, although methyl mercury, the most harmful compound, is not detected (Table 5). The level of Hg in the ashes is very low and may cause no environmental problems.

A commercial unit to remove 95% of Hg in the combustion gas went into operation in 1988 for a 300 ton/day incinerator. The unit applies sodium scrubbing using a liquor added with about 100 mg/liter of NaClO.

The waste liquors leaving wet scrubbers usually contain Hg at

TABLE 5
Mercury in Waste Gas and Ashes from MSW Incinerators[a]

	Waste gas (*μg/N m^3*)	*Fly ash* (*mg/kg*)	*Bottom ash* (*mg/kg*)
Total mercury	10·9–55·3	0·27–6·6	0·18–0·28
Methyl mercury	ND[a]	ND–0·0023	ND

[a] By Japan Environment Agency.
[b] Not detected.

1–10 mg/liter. The liquor is treated with a chelating compound to precipitate and separate out the Hg and to reduce Hg in the wastewater below 0·005 mg/liter to meet the effluent standard. A commercial unit was constructed recently to recover more than 90% of the Hg in the scrubber liquor as metallic mercury.

Dioxines and Dibenzofurans

Minor amounts of polychlorinated organic compounds such as dioxines and dibenzofurans are formed during the combustion of MSW and the cooling of the gas and are contained in the waste gas as well as in the ashes (Table 6). Polychlorinated dibenzoparadioxines (PCDDs) and polychlorinated dibenzofurans (PCDFs) are detectable in the gas and ashes but the most hazardous, 2,3,7,8-tetrachlorinated dibenzoparadioxine (2378-TCDD), is very minor or not detectable. Further studies on the compounds are needed although no appreciable environmental problems may be caused by these compounds.

Large-scale tests indicated that from 1 ton of MSW, 3–10 mg of PCDDs and 1–3 mg of PCDFs formed when the combustion gas contained 6–7% O_2 while both decreased to 0·3–1 mg as O_2 increased to 12%. The increase in O_2, however, is not desirable because of the NO_x increase. The tests also showed that 90–95% of the compounds were removed by lime slurry spray and an ESP, while more than 99% of them were removed by lime slurry spray and a fabric filter giving emission concentrations below 1 ng (10^{-9} g/N m^3).

Gas Cleaning for Sewage Treatment Sludge Incinerators

The flue gas from sewage treatment sludge incinerators contains particulates, SO_x, NO_x and offensive odor. The gas around 250°C is usually subjected to sodium scrubbing. In Japan about 20 SCR plants have been constructed to treat the tail gas using the system shown in

TABLE 6
Polychlorinated Organic Compounds in Waste Gas and Ashes From MSW Incinerators (by Japan Environment Agency)

	Waste gas (ng/N m^3)	*Fly ash (μg/kg)*	*Bottom ash (μg/kg)*
PCDDs	101–924	81–1251	9·3–43·1
2378-TCDD	0·28–1·20	ND[a]–5·0	ND
PCDFs	231–762	22·4–1510	0·62–589

[a] Not detected.

Fig. 2(b). The gas heated at 350–400°C is treated by a special catalyst with 1 mol NH_3 to 1 mol NO_x to remove about 90% of both NO_x and offensive odor. The odorous compounds such as mercaptan are decomposed in the reactor by oxidation.

CO_2 EMISSION AND CONTROL

CO_2 Increase Caused by Flue Gas Cleaning

The levels of CO_2 in ambient air have increased significantly with fossil fuel consumption and may have serious adverse effects on the environment. There may be no practicable way to remove CO_2 from combustion gases and so all kinds of efforts should be made to depress CO_2 emissions.

FGD consumes materials and energy and thus increases CO_2 production. Limestone reacts with SO_2 to release CO_2. Lime scrubbing further increases CO_2 because lime production from limestone gives CO_2 derived from both limestone and the fuel. The production of absorbents such as ammonia and sodium hydroxide also produces CO_2. The power consumption by FGD causes an increase in fuel use of 1–2%. Among FGD processes, limestone scrubbing causes a relatively small CO_2 increase, and SCR causes an even smaller CO_2 increase because of the low ammonia and power consumptions. Moderate combustion modification based on excess air reduction can depress both fuel consumption and NO_x formation.

CO_2 Emissions by Electric Power Generation

Table 7 shows estimated amounts of CO_2 produced to generate 1 kWh of electricity by fossil fuels. The emissions include those by the fuel and flue gas cleaning but exclude the emissions due to the mining and transportation of the fuels, the construction of power plants, etc.

Among the fossil fuels, natural gas gives the least CO_2; oil produces about 40% more than gas, while coal gives even more. Coal with limestone–gypsum FGD and SCR may give nearly twice as much CO_2 as does natural gas by combined cycle.

Oil and gas should be used for high-efficiency power generation. NO_x production may be increased by high-temperature combustion used to improve efficiency but this can be reduced with SCR, giving overall advantage. On the other hand, coal gasification and liquefaction increase CO_2 because of the high energy consumption which may not be compensated by improved efficiency by the combined cycle.

TABLE 7
CO_2 Emissions by Electric Power Generation

Fuel	*Power generation*	*Efficiciency (%)*[a]	*Gas cleaning method*	*CO_2 emissions*	
				(kg/kWh)	*Ratio*
Natural gas	Boiler	39	None	0·46	100
Natural gas	CC[b]	44	None	0·41	89
Natural gas	CC	44	SCR	0·414	90
Natural gas	CC	47[c]	SCR	0·386	84
Oil, high S	Boiler	40	None	0·66	143
Oil, high S	Boiler	40	FGD	0·68	148
Oil, high S	Boiler	40	SCR + FGD	0·684	149
Oil, low S	CC	44	None	0·60	130
Oil, low s	CC	44	SCR	0·608	132
Coal (2·5% S)	Boiler	40	None	0·75	163
Coal	Boiler	40	FGD	0·776	169
Coal	Boiler	40	SCR + FGD	0·78	170

[a] Based on high heat value, under optimum condition.
[b] Combined cycle.
[c] Future possibility.

MSW and sewage sludge incineration in Japan may produce CO_2 equivalent to 10 million tons/year of coal. Although power is generated at large MSW incineration plants, the heat efficiency is only 10–20%. It would be better to produce a compost from garbage and sludge, and to recycle papers and plastics as far as possible.

REFERENCES

Ando, J. & Kaplan, N. (1988). Recent developments in SO_2 and NO_x abatement technology in Japan and China. Presented at the First Combined and Dry SO_2 Control Symposium, St. Louis, October.

Delleney, R. D. & Dickerman, J. C. (1988). FGD research and development status. Presented at the First Combined and Dry SO_2 Control Symposium, St. Louis, October.

Maknsi, J. & Ellison, W. (1988). Worldwide progress in the utilization of by-product gypsum. Presented at Seminar on New Frontiers for By-product Gypsum. Mississuauga, Ontario, November.

8

Wastewater Treatment Technologies

M. Rebhun and N. Galil

Environmental and Water Resources Engineering, Technion—Israel Institute of Technology, Faculty of Civil Engineering, Technion City, IL-32000 Haifa, Israel

INTRODUCTION

The development of wastewater treatment processes is related to treatment objectives, which are in turn dictated by wastewater characteristics and effluent quality requirements or standards. Although major constituents of municipal wastewater have not changed during the 80 or 100 years of modern wastewater systems, many new constituents and chemicals have appeared in the last three decades in municipal wastewater, due to technological advances in new products, changes and modernization in household habits and increased urbanization. Of particular significance was the appearance of synthetic detergents and associated washing formulations with widespread use of phosphates and the many synthetic organic chemicals widely used for a variety of household purposes. The major sources of these 'new' problematic constituents is industrial wastewater discharged into public sewers and reaching the wastewater treatment systems.

Effluent quality requirements are related to quality criteria of receiving waters into which the effluent is discharged or to quality demands of consumers of reclaimed wastewater in cases where the effluent is re-used. The increasing load of pollutants on receiving waters and the ongoing deterioration in the quality of many water bodies (rivers, lakes, groundwater) have caused the introduction of higher quality effluent standards and inclusion of new parameters in these standards.

The combination of wastewater characteristics with the effluent standard requirement determines the treatment objective. Treatment technologies are developed and introduced so as to meet these objectives.

CHANGING TREATMENT OBJECTIVES

Historical Objectives

The historical quality objectives from the early days of modern wastewater treatment were as follows:

(1) Prevention of environmental nuisances and of negative aesthetic effect, such as obnoxious odors, discoloration, deposits, etc.
(2) Maintenance of sufficient levels of dissolved oxygen in receiving waters (rivers, lakes). Thus, concentrations of oxygen-demanding components in effluents should be limited and meet the needs of the recipient.
(3) Prevention of danger to public health. In some instances disinfection of effluent may be required, the effluent quality should be such as to enable efficient disinfection.

The treatment objectives derived from those quality objectives were: (1) removal of suspended solids; (2) removal of BOD; (3) in some cases oxidation of ammonia (nitrification)—considered as necessary because of ammonia's toxicity to fish and its oxygen demands exerted upon water bodies.

Present-Day Objectives

The period since World War II can be characterized by a rapid increase in urbanization with great population increase in large cities and metropolitan areas associated with new living habits, by a rapid and intensive industrial growth and by a tremendous development of new synthetic products of diverse nature.

These developments affected the wastewater situation significantly. The total amount of wastewater produced rose drastically and as a result the mass load of the historical pollutants suspended solids and BOD increased even in effluent meeting the 'old' standards.

Of particular significance was the appearance of high concentrations of phosphates due to widespread use of synthetic detergent formulations with high phosphate content. This high load of phosphates and nitrogen compounds discharged with effluents resulted in stimulation of excessive algal growth in receiving waters and the risk of man-induced eutrophication. It triggered a new quality objective and the development of new technologies: nitrogen and phosphorus nutrient control. The need to deal with nitrogen control stems also from the increasing nitrate content in various waters and its associated health significance.

The tremendous developments of new synthetic organic chemicals (SOC) and production increase in industries using heavy metals resulted in the appearance of a variety of synthetic organics and of increasing concentrations of heavy metals in wastewaters. Many of these compounds exert a toxic or inhibitory effect on treatment processes. Most are recalcitrant, remaining in treatment plant effluents and having adverse ecological impacts with high bioconcentration factors in the food chain. Many of the SOC are of health-related significance, producing effects such as carcinogenicity, mutagenicity and conventional toxicity. In recent years these compounds are referred to as 'hazardous compounds' with close attention being given to and significant resources allocated for their study and control.

In view of the situation described, present-day effluent quality objectives include:

(a) Much lower concentration levels of the 'historical' parameters—the suspended solids and BOD—are nowadays required to cope with the higher total volume of wastewater produced.
(b) Nutrients control in effluent. Low concentrations of phosphate and nitrogen compounds are now major effluent quality objectives.
(c) More and more often organic micropollutants (SOC) have to be controlled and their content in effluents limited to below levels causing ecological effects or health-related problems in the receiving water.

When effluent is to be reclaimed and re-used, appropriate quality objectives are derived according to use: irrigation, industry or general re-use (after mixing with other sources) even for public supplies. For irrigation, mainly public health quality objectives dominate, while presence of nutrients may be desirable. For industrial use and complete renovation far-reaching quality objectives are involved. For most re-use purposes preparation of the effluent for efficient disinfection is necessary.

The present-day treatment technology objectives can be summarized as follows:

(1) Removal of BOD and suspended solids to very low levels.
(2) Removal of P and N nutrients.
(3) Adaptation of treatment processes to the presence of toxic or inhibitory pollutants.
(4) Removal of recalcitrant micropollutants having adverse ecological and health effects.
(5) Preparation of effluent for efficient disinfection wherever disinfection required, avoiding disinfection products with negative effects on the aquatic environment.

OVERALL TREATMENT PROCESS AND PERFORMANCE

The overall process flow sheet and the performance of its component unit processes are shown in Figs 1 and 2. The process flow scheme can be divided into sections, each fulfilling a given group of treatment objectives.

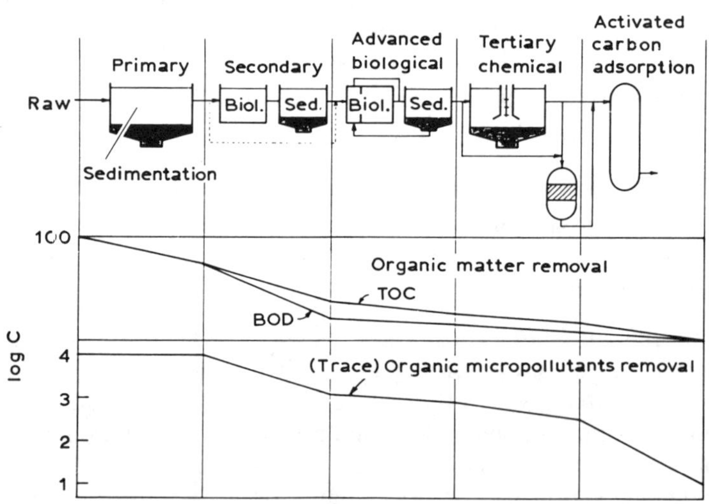

FIG. 1. Overall treatment process flow scheme and performance profiles for organic matter removal and organic micropollutants removal.

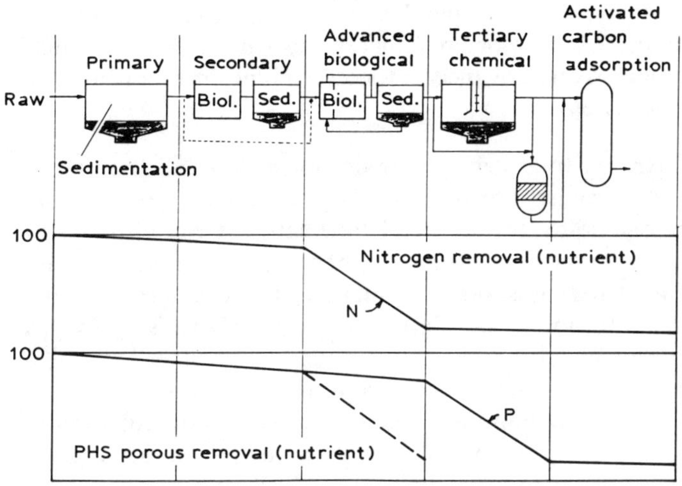

FIG. 2. Overall treatment process flow scheme and performance profiles for phosphorus and nitrogen nutrient removal.

(1) The 'historical' treatment objectives—reduction of suspended solids and BOD—are achieved through the primary sedimentation and secondary biological treatment processes.
(2) Nutrients removal is accomplished by: (a) advanced biological processes for nitrogenous matter removal by nitrification–denitrification and recently also attempted for biological phosphorus removal; and (b) chemical treatment utilized for phosphate precipitation–flocculation followed by sedimentation–clarification.
(3) The reduction of suspended solids and BOD to concentrations below those obtained by secondary treatment is accomplished either by tertiary filtration with or without chemical flocculation or concurrently with the tertiary chemical treatment for phosphate removal.
(4) The removal of trace heavy metals is also achieved in the chemical treatment. Thus chemical treatment followed by sedimentation–clarification and filtration can accomplish and fulfil a combination of treatment objectives.
(5) Removal of recalcitrant organic micropollutants is achieved through adsorption on activated carbon. This process, if included, is usually the last except for disinfection.

(6) Disinfection is applied as needed after any of the listed processes, its efficiency increasing when applied to effluents of progressively advanced processes going from left to right on the flow scheme.

Most existing treatment plants do not include all the unit processes shown. Most plants are of secondary treatment type with increasing numbers extending to advanced biotreatment for nitrogen reduction and some chemical treatment either separately or combined with biotreatment for phosphorus reduction. There is also a clear increase in tertiary filtration in developed countries where 'sensitive' receiving waters are involved.

For specific re-use purposes additional processes are sometimes applied, such as membrane processes for partial desalination to reduce either total salinity or specific mineral salts 'picked-up' during the use of the water.

In the next paragraphs, each of the unit processes mentioned above will be discussed.

BIOLOGICAL TREATMENT PROCESSES

Biotreatment processes are the mostly used in wastewater treatment. The principles, the mass and energy flow in bioprocesses, are shown schematically in Fig. 3. Most of the substrate is used for biosynthesis of cells—biomass, and the remainder oxidized to produce energy which is then used for the biosynthesis (cell growth). Some of the cells (the biomass) are also oxidized in the process of endogenous respiration. The resulting products of the process are oxidation products and stable organic matter, which leaves with the effluent and biomass or biosludge which has to be dealt with and disposed.

Three major types of biomass are encountered in wastewater treatment bioprocesses.

(1) bioflocculated suspended growth;
(2) fixed film growth; and
(3) dispersed algal cells (photosynthetic).

Major reactor types corresponding to each type of biomass are shown in Figs 4–6.

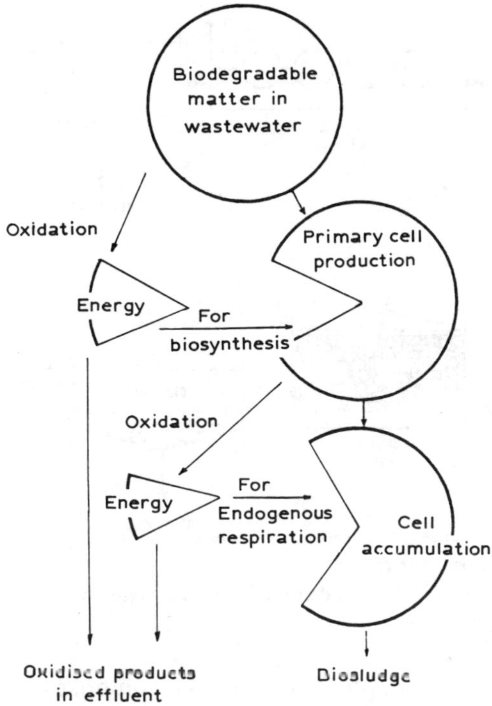

FIG. 3. Schematic of mass and energy flow in biotreatment.

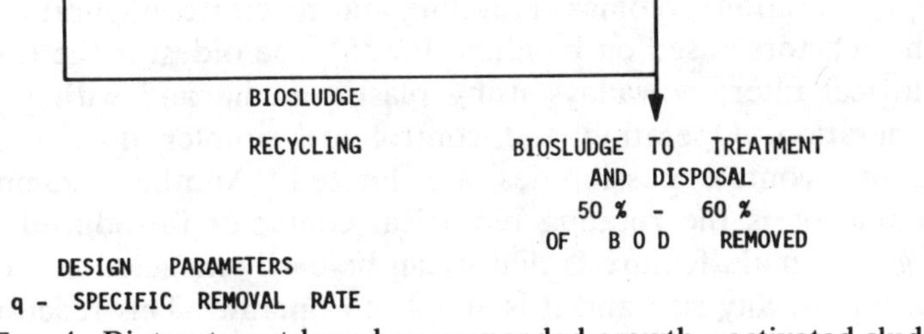

FIG. 4. Biotreatment based on suspended growth—activated sludge.

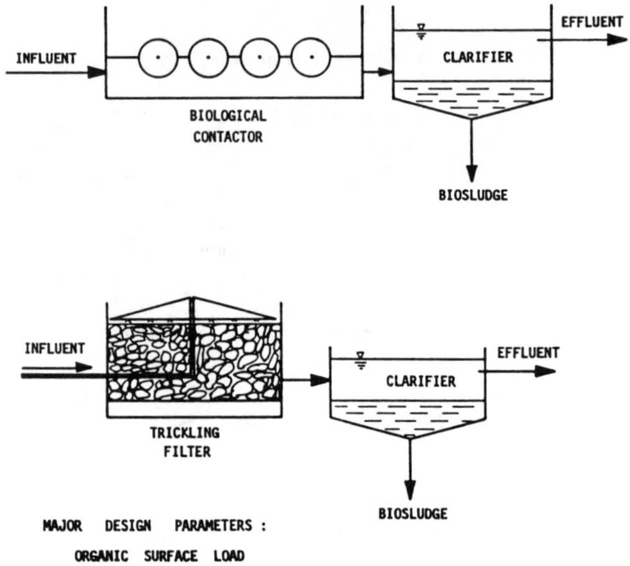

FIG. 5. Biotreatment based on fixed films.

The activated sludge process (Fig. 4) comprises a reactor with suspended biomass, biomass settling in a secondary clarifier and biomass recycle. The cells (biomass) are in a flocculated form—the process mechanism includes bioflocculation—and operation should aim for good biomass settling properties. Basic design and operational parameters are organic loading per unit biomass and biomass (sludge) age. Of all biotreatment processes this process is the easiest to control by regulating the major process parameters such as biomass age and recycle. It does require skilled operators and close operation control. There are many modification of this process with regard to methods of air supply (aeration), biomass recycling and reactor configurations.

Of the reactors based on biofilms (Fig. 5), the oldest is the trickling or biological filter, nowadays using plastic media and with possible forced aeration. Operation and control are simpler than activated sludge, but control possibilities are limited. Another example of biofilm reactor is the rotating biological contactor introduced 10–15 years ago. Its main feature is that it can be easily installed in modular units of almost any size and it is simple to operate. This reactor type has been constantly undergoing many technical modifications to

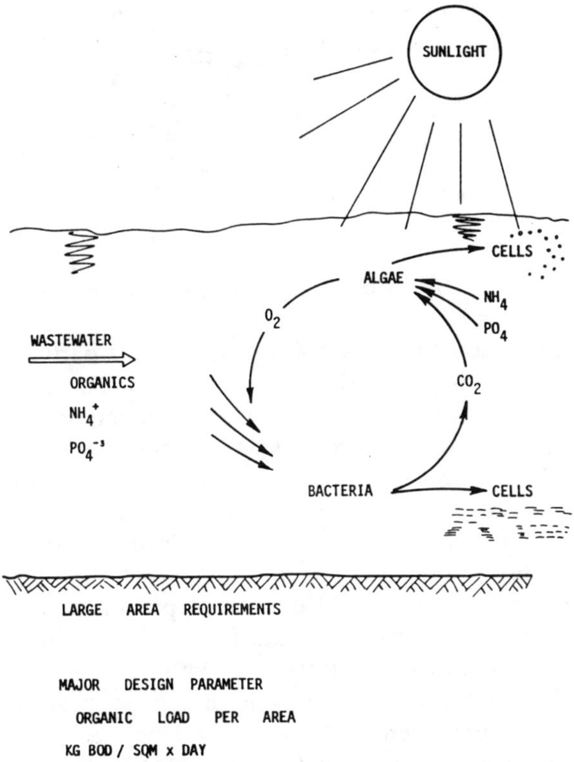

FIG. 6. Biotreatment based on algae stabilization (oxidation) ponds.

overcome mechanical problems in its operation. In the last decade the biofilm area has been receiving considerable attention in research on its fundamental behavior. It is believed, therefore, that this may lead to significant practical development of this process in the future.

Low-cost and simple-to-operate treatment processes are oxidation (or stabilization) ponds (Fig. 6) based mainly on symbiosis of bacteria and algae, the driving force being photosynthesis and algal cell production, with accompanying oxygen production needed for the process. This is a suspended biomass (growth) type process but the cells are not flocculated, are mostly planktonic and not in a separable state. Because the energy source is solar radiation and so the process is area-intensive, this method is particularly suited for warm climates and where low-cost land is available. No mechanical devices or any power

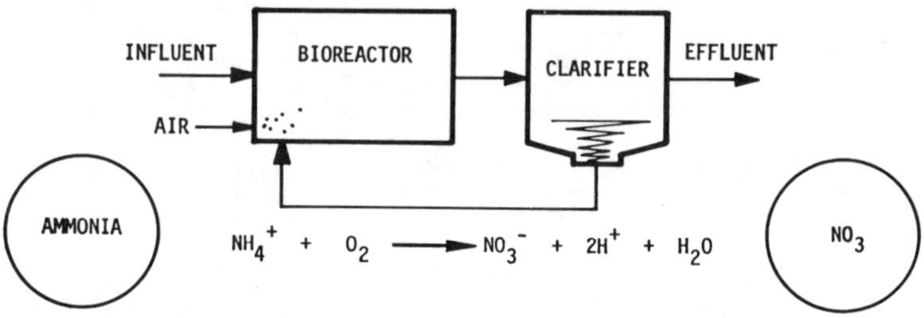

FIG. 7. Bio-oxidation of ammonia—nitrification.

source are necessary to operate the process. It is particularly suitable in tropical, developing countries and rural areas. The effluent contains algal cells in high concentration (high suspended solids); these are relatively stable organic particulates and do not exert oxygen demand in daylight. In systems arranged as several ponds (reactors) in series, low algal concentrations can be achieved due to die-off and predator development consuming the algae and clarifying the water. This obviously requires long residence times and 7–20 days residence times are common for such systems. Removal of pathogens in such systems is very efficient—an important feature for tropical regions.

In all of the bioprocesses described, with reasonably good operation, the soluble fraction of degradable organic, the BOD, is removed efficiently and the effluent soluble BOD levels are low. The main problem affecting effluent quality in practice is the separation efficiency of the biomass in the secondary settling tanks (clarifier). The settlability of the biomass and the behavior of the secondary clarifier are the critical factors in design and operation of these systems.

Oxidation of ammonia is often required before discharging into the receiving water or re-use. This can be accomplished by biological nitrification (Fig. 7) or by some of the processes described. The best and most reliable for this purpose is the activated sludge process and its modification. Long cell residence times (sludge age) and sufficient dissolved oxygen levels are needed for efficient nitrification. The process is sensitive to many inhibitory compounds.

NUTRIENT REMOVAL

Chemical Processes

The need to remove N and P gave rise to 'new' technologies in the wastewater area. The removal of phosphates was the first and main

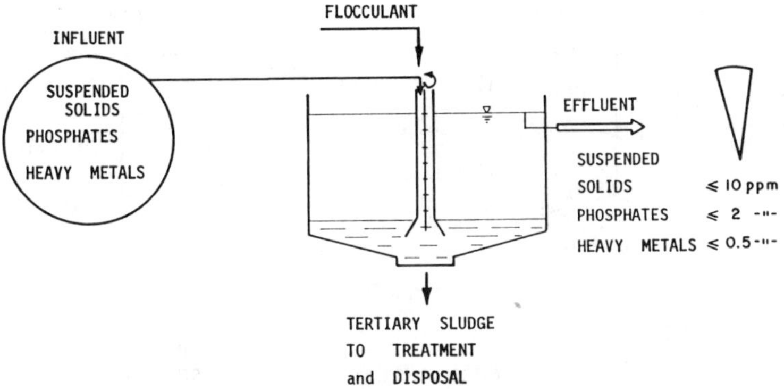

FIG. 8. Tertiary chemical treatment and sedimentation–clarification.

target. This could be accomplished efficiently by chemical precipitation followed by clarification through settling and sometimes filtration. A schematic of a typical process is shown in Fig. 8. Originally, the process was introduced as a tertiary stage and is still mostly used in this way. Chemicals applied are aluminium and ferric salts or lime. These chemicals also act as flocculants, and so with phosphates precipitation, flocculation–clarification is also achieved. In the initial stages of development of nutrient control technologies nitrogen removal was intended by ammonia stripping. The overall process was a combined phosphorus and nitrogen removal method where lime was used for P precipitation and raising the pH to convert ammonium NH_4^+ to volatile ammonia NH_3, the latter to be stripped off in a forced-draft stripping tower (Fig. 9). Although several full-scale plants were erected, due to many technical problems inherent in this method such as excessive scaling and reduced removal efficiency, this process has not been adopted. The common process applied nowadays for nitrogen control is biological nitrification–denitrification which will be discussed later. Lime treatment at tertiary stage is used in a number of plants for phosphate control, giving very low P residuals. Lime treatment provides for efficient heavy metal removal, high bacteria removal, and softening factors very important in treatment for re-use.

Biological Processes for Nutrient Control

The process that was adopted and is widely used at present for nitrogen control is biological nitrification–denitrification. One system employs two consecutive reactors with two separate settlers and

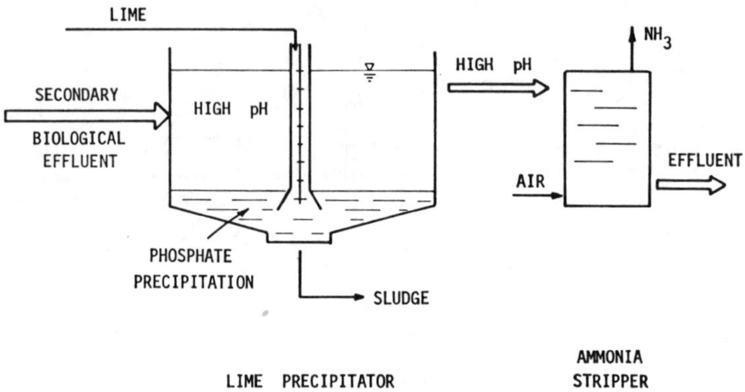

FIG. 9. Early physico-chemical process flow sheets for P and N removal.

separate biosludges, in which case a readily available organic carbon source has to be added to the denitrification stage. An alternative is a single-sludge system with one settler and the bioreactor divided into an aerated section and an anoxic section into which the liquid from the aerated section is recycled. The carbon source is the organics in the raw sewage. The scheme can be seen in Fig. 10. The level of nitrate in the effluent is regulated by the recycle ratio.

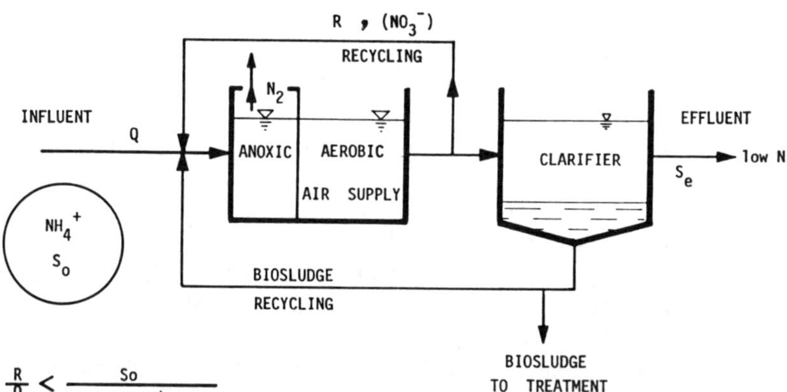

$$\frac{R}{Q} < \frac{So}{2.3 * (NH_4^+) - S_o}$$

$$(NO_3) = \frac{NH_4}{\frac{R}{Q} + 1}$$

FIG. 10. Advanced biotreatment for nitrogen removal.

This is the most common, reliable and cost-effective nitrogen removal process employed at present in the field. Attempts have been made for more than a decade to incorporate biological phosphorus removal in this process by adding an anaerobic section, in addition to the aerobic and anoxic sections. The residence in the anaerobic section should cause an enhancement of P uptake in the aerobic section. These attempts are part of the overall tendency to return to bioprocesses as the only or main treatment for all constituents.

TREATMENT OF TOXIC AND HAZARDOUS ORGANICS

Secondary effluents (even after tertiary chemical flocculation–clarification) contain recalcitrant soluble organic compounds which either originate in the raw wastewater or are formed in the biotreatment process. These organics are responsible for the color and odor of effluents, are fouling agents in water treatment processes and complexing agents in waters to which effluents are discharged. But most important, this group contains environmentally and health-related hazardous organic compounds (organic micropollutants) of toxic, carcinogenic, mutagenic and other damaging effects. With increasing problems and awareness regarding hazardous compounds, effluent quality is evaluated not only by the common parameters of BOD, SS, nutrients, etc., but more and more frequently by their toxic acute or long-range effects.

Therefore more attention will be given to these compounds in the overall treatment scheme.

Adsorptive Processes

The removal of such organics can be accomplished by adsorption on activated carbon. Most of these organics are of hydrophobic nature and, therefore, have favorable adsorptive equilibria towards activated carbon surfaces.

Activated carbon adsorption is an advanced and costly process. It can be justified economically when effluents containing such organics have to be disposed into 'sensitive' recipients with high-quality standards and low dilution ratios. Adsorption is often employed in water renovation treatment systems for wide re-use objectives, such as recharge to high-quality aquifers, industrial production re-use and always for public supply re-use. The most common configuration of

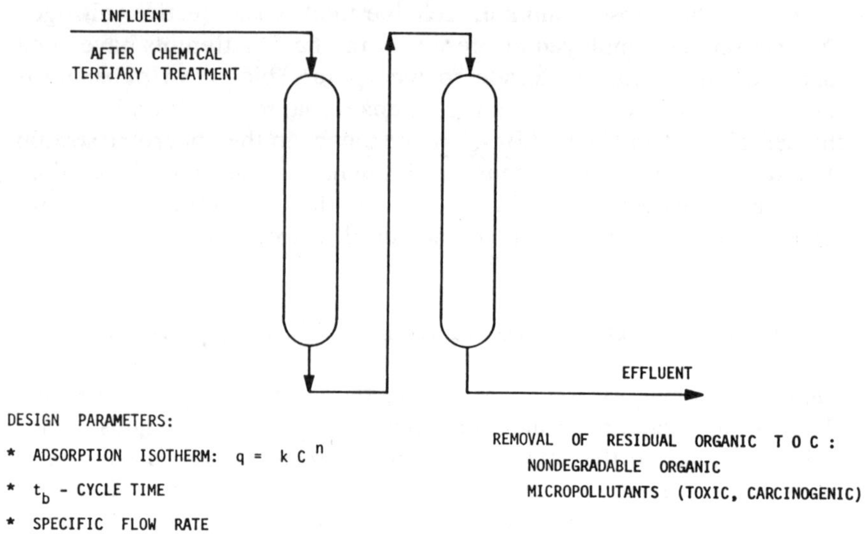

FIG. 11. Advanced treatment by adsorption on activated carbon.

this process is the use of granular activated carbon in columnar reactors (Fig. 11). The major design parameter is the adsorptive capacity determined by the adsorption isotherm. This capacity is a function of the concentration of the organics to be removed. In effluents these concentration are low and thus working capacities of carbon for effluents are much lower than for adsorptive processes in industrial operations. Also, problems of competitive adsorption are encountered and have to be taken into consideration. Effluents are multisolute systems with bulk organics taking up a great part of the capacity and out-competing the trace organics. Adsorption is an exhaustive process and the carbon has to be replaced and regenerated. On-site carbon regeneration is a serious operation and can be cost-effective in relatively large installations.

Often biological growth (films) occurs in carbon columns treating effluents. Consequently, a combination of adsorptive and biological processes takes place in such situations. With increasing awareness of hazardous organic compounds, adsorption processes may find wider use in future treatment schemes.

Chemical Oxidative Processes

A possible alternative to adsorption as a process to remove recalcitrant organics is chemical oxidation. Research is being done to develop oxidative processes to 'detoxify' such compounds by oxidation. It would be preferred if such processes would lead to complete oxidation but most often they lead to the formation of new oxidized compounds of no or unknown toxic effects. Addition of chemical oxidants is a simpler operation than active carbon adsorption, but the question of the nature and effects of the oxidation products remains to be resolved.

Chemical Oxidation Followed by Biodegradation

An attractive research and development activity is the oxidation of recalcitrant organics into products that are biodegradable. Oxidation of organics usually results in increase in hydrophilic groups, leading to enhanced bio-availability. The most popular oxidant for this purpose is ozone, although chlordioxide has also been considered. The biodegradation processes that follow are usually in biofilm reactors on granular media.

In treatment of industrial hazardous organic wastewater (high organic concentrations) the biodegradation process, following the chemical oxidation preparatory process, could be in suspended biomass reactors with recycle (the activated sludge type).

New Biotechnologies for Hazardous Wastewaters

The tendency to prefer biological treatment processes can also be seen in the hazardous waste areas. Intensive research and development is proceeding to develop bioprocesses using co-substrates and/or co-enzymes to enhance the biodegradability of hazardous organic compounds in industrial wastewater.

NON-CONVENTIONAL PROCESS CONFIGURATION AND SEQUENCES

'PCT Only'

The introduction of physico-chemical treatment (PCT) processes for nutrient control in the 1960s, brought about a short 'era' of attempts and proposals to use physico-chemical treatment as the only treatment for wastewater. A typical flow scheme can be seen in Fig. 12, consisting of chemical flocculation at the primary stage followed by possible ammonia stripping, granular media filtration and activated carbon adsorption for the removal of organic matter. The attractiveness of PCT was that it is much less sensitive than biological processes

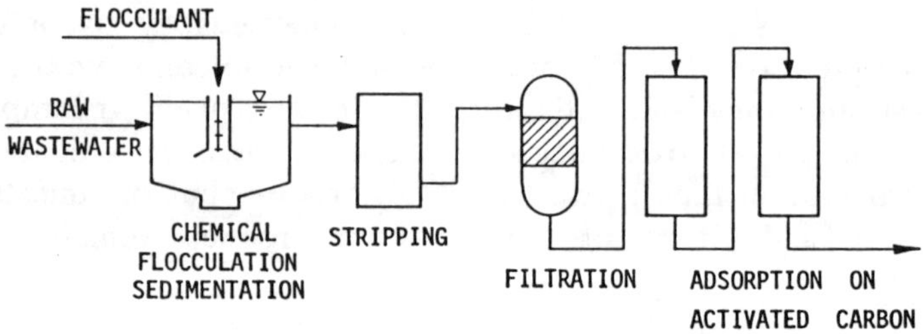

FIG. 12. Typical flow scheme using physico-chemical treatment alone.

to fluctuations and to environmental factors and that it is more easily controlled. In practice, however, biological treatment could not be eliminated. The bulk of soluble organic matter in raw municipal wastewater is poorly adsorbable on activated carbon, particularly the biodegradable organics, because of their highly hydrophilic nature and low molecular size. Thus for the bulk of soluble organics in raw sewage carbon adsorption is thermodynamically unfavorable while biological treatment is favorable.

The chemical treatment at primary stage, however, removes more than 50% of BOD, phosphates, most heavy metals, and other objectionable constituents, and is thus being a good pretreatment before biological processes.

Primary Chemical Treatment Followed by Biological Treatment

While the 'PCT' only scheme proved to be impractical, the chemical treatment at primary stage produced a primary effluent superior in quality to primary effluents from plain sedimentation. The reduction in BOD is close to or exceeds 50% vs. 30–35% reduction in plain primary treatment, all of it remaining in soluble form. This means lower loadings on the biological units and a more readily available substrate resulting in a higher process rate and lower residence times. The primary chemical treatment removes many toxic elements such as most heavy metals and other agents interfering with biological treatment. This contributes also to increased process rates and much better performance of the bioprocess, particularly the nitrification process. Obviously, the tertiary chemical treatment for P removal is not needed in such configurations. This configuration produces somewhat higher amounts of primary sludge (chemical type) as compared to plain sedimentation. There are several large-scale plants employing this process sequence.

With more hazardous wastes reaching the municipal wastewater systems, this process sequence with primary chemical treatment may become useful for removal of hazardous compounds before biological treatment to prevent inhibition effects on the bioprocess.

SLUDGE TREATMENT

Most of the pollutants that are removed in wastewater treatment, particularly many of the objectionable pollutants, end up in the various sludges that are formed in the unit processes and operations comprising the overall treatment process. The sludge flow and treatment scheme is shown in Figs 13 and 14.

These sludges have to be concentrated, if necessary stabilized, dewatered and disposed of in a way that does not cause environmental damage or water pollution. The primary and secondary treatment sludges are biologically stabilized, concentrated and dewatered. The chemical sludges are concentrated by thickening and then dewatered. Dewatering is accomplished by cake filtration using vacuum or belt filter processes. In the last decade belt filters have been most popular.

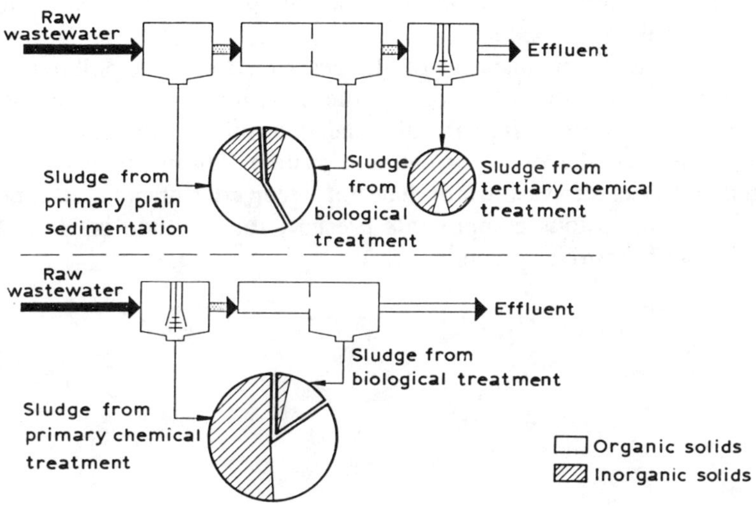

FIG. 13. Sludge production in wastewater treatment.

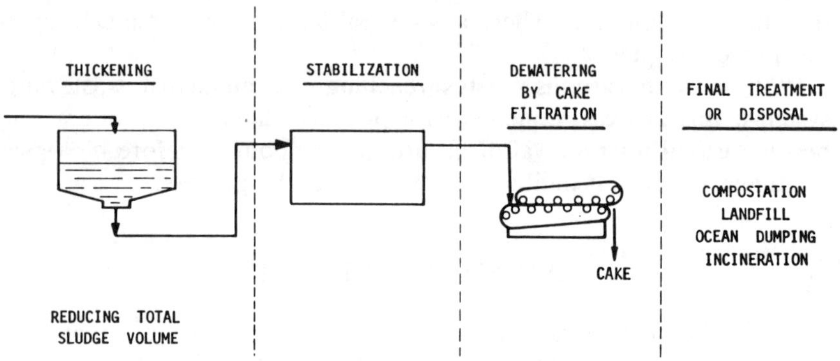

FIG. 14. General flow sheet of sludge treatment.

Flocculant conditioners are used to improve thickening and dewatering rates and properties. The final product should be a solid agglomerate that can be easily handled and disposed of as a solid waste. It should be remembered that these products of sludge dewatering often contain hazardous compounds in high concentrations. Their disposal into 'depositories' should be carefully planned so as not to cause contamination of groundwaters, soils and environmental resources in general, by leaching, extraction and dissolution of these contaminants. In the case of sludges containing hazardous organic pollutants, incineration with complete—often catalytic—oxidation is being considered.

The treatment and disposal of sludges which contain pollutants in concentrated form should be one of the most important operations in wastewater treatment. Its safe disposal determines the real environmental value of the overall treatment system. Unfortunately, in the past this aspect of wastewater has not received sufficient attention. With more hazardous components present, this attitude is obviously changing in a positive direction.

9

Water Pollution Control with Special Reference to Wastewater Treatment

HELLMUT FLECKSEDER

TU Wien, Institut für Wassergüte und Landschaftswasserbau, Karlsplatz 13, A-1040, Vienna, Austria

INTRODUCTION

Human insights and views are shaped by the traditions in which we live. The author trained in a Central European civil engineering school, but has had periods of work elsewhere. He mainly works in the applied field in the area of wastewater treatment and in counselling the Austrian authorities in the field of water pollution control.

This paper puts forward the view that wastewater treatment is only one facet of the many-sided coin of water pollution control. In this approach wastewater treatment is placed in the context of a scheme consisting of formation of wastewater, sewage, treatment itself and the discharge of treated wastewaters to recipients (groundwaters, rivers, lakes, estuaries, inland seas, oceans) as well as of residues to land or soil. This approach clarifies the longer ranging position of wastewater treatment. The atmosphere is excluded from this presentation for the sake of simplicity. The starting points for this paper were applied problems; it is therefore not strictly scientifically oriented.

A SIMPLE SYSTEM REPRESENTING WATER POLLUTION CONTROL

Water pollution control is understood to be a general human behaviour which aims to protect aquatic recipients and the habitat they offer to indigenous species from inputs caused by human misbehaviour. This general human behaviour is controlled by administrative rules, by application of technologies, but also by direct insight of the individual human being into the necessities of his individual behaviour towards aquatic recipients. Water pollution control is part of protection of the environment, but mankind lives (and has to live) by utilizing the same environment.

What is wastewater? For the purpose of this paper in connection with water pollution control, wastewaters are all those waters which require some type of treatment before they can be safely discharged to a recipient. The term 'safely' in this context implies that the conditions in the recipient in relation to the quantity and quality of the type of wastewater to be discharged determine whether or not treatment has to take place. From this, sewage or industrial wastewaters are wastewaters according to this definition, but drained rainfall from paved urban surfaces is, if sewered separately and according to our present knowledge, in many applied cases (but not in every case) not a wastewater according to this definition. From a hygiene point of view, however, drained rainfall from paved urban surfaces is an actual wastewater, as it cannot be put to unrestricted common use.

As shown in Fig. 1, water pollution control occurs along the path from formation of wastewater via sewerage to wastewater treatment and from there on to the aquatic recipients, but also from land/soil to the aquatic recipients. However, this simple system representing the main streams of water pollution control is not yet complete, as the formation of wastewater does not occur out of nothing; it arises from the usage of matter—generally originating from land/soil—brought into contact with water. Water also cycles back from the recipients to its use. Matter taken out of wastewaters by treatment has to have an outlet back to a 'solid recipient', i.e. land/soil. This in turn means that reasoning about water pollution control automatically invokes reasoning about environmental protection in general. The system presented also demonstrates that the feedbacks which determine the allowable flow of matter (and energy) through the system have to start from both recipients and pass back through all compartments of the system mentioned.

What do the individual compartments of this simple system look like?

The compartment 'formation of wastewater' comprises the transfer of matter into water plus all the measures of hindering such a transfer. In most households in Europe, the situation is at present simple: the

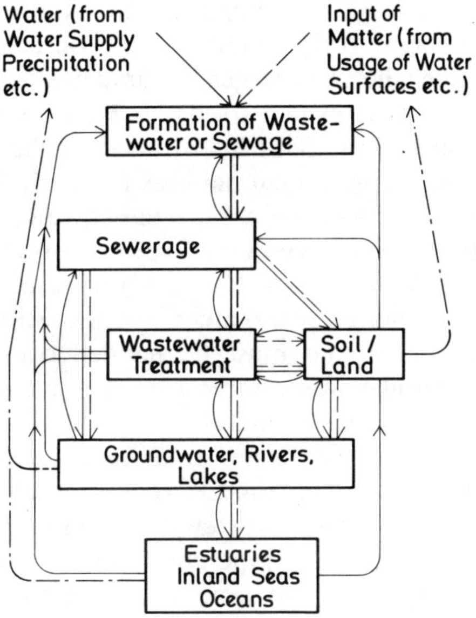

FIG. 1. A simple system demonstrating the compartments, flows of matter and feedback in water pollution control.

end-products of human metabolism plus all transfer of matter from washing, form sewage. In industry and trade, the tendency in development both in processing technology as well as in pretreatment is to lose as small an amount of waste as possible into wastewater.

In the early days of wastewater treatment, pretreatment requirements primarily considered the link to reliability of wastewater treatment, but not so much the links to the recipients. In the late 1970s, the pretreatment requirements became much more stringent, and the more so in recent years.

If a product to be marketed, e.g. fibreboard, does not need to be washed intensively, the tendency has been to convert any wet processes in production into dry ones; so that any make-up water necessary can be lost in the form of steam. However, substitution of every wet process by a dry one has not been possible for reasons of product quality. In any process in which matter has to be washed out in order to produce clean end-products, a wastewater will be formed, whether it is to be treated—at the pretreatment stage or at the end-of-pipe stage. The general application of these developments in process technology and pretreatment depends on economic criteria.

In storm drainage, the present general practice is to drain all paved surfaces, but there is a trend to reduce the amount of these, thus reducing the amount of rainfall to be drained artificially. Rainfall and its disposal by drainage in human habitats is closely linked to the type of sewerage, which leads to the compartment 'sewerage'. In drainage practice in Central Europe, combined sewers are chosen more often than separate ones. Both drainage and sewerage originated primarily for hygienic reasons, but they also arose because wastewater cannot be treated easily at source and must be handed over to recipients of sufficient size. Sewerage was always part of developed civilizations, e.g. the ancient Greeks and Romans.

The compartment of 'wastewater treatment' in the context of this simple general system is understood to represent a technology applied to purify sewage or the overall wastewater from an industry; it is thus a technology applied at end-of-pipe and is distinct from pretreatment which takes care of specific waste constituents and which in this system is part of the compartment 'formation of wastewater'. The wastewaters in question usually contain biodegradable carbon (determined by COD_b, BOD_5, etc.), nitrogenous and phosphorous species, and are amenable to conversion and removal by biological or biological–chemical treatment. The scientific study and application of biological wastewater treatment dates back to Frankland, with intermittent filtration (Dunbar, 1907), and to Ardern & Lockett (1914), with the activated sludge process (e.g. v.d. Emde, 1984), who all lived in the UK.

The task of wastewater treatment is to split sewage/wastewater into two streams: (a) the treated wastewater (recipient: the aquatic environment); (b) the residues from treatment (recipient: land/soil). It has been known since 1886 that biological processes allow the removal of biodegradable carbon and the oxidation of ammonium to nitrate. Our present knowledge tells us that the oxidation from ammonium to nitrate also assures the oxidation of most of the organic compounds present in the respective wastewater. Organic compounds which are not oxidized under the conditions prevailing in nitrifying plants should not, if possible, enter wastewater in its formation.

Simultaneous precipitation of phosphorus (Thomas, 1962), single-sludge nitrification–denitrification (Matsché, 1972) and biological phosphorus removal (Barnard, 1975) are more recent techniques.

For decades, the aquatic environment considered by the sanitary engineer was the adjacent groundwater, the nearest creek, river or lake. Formulas for the assimilative capacities of rivers were studied quite early (e.g. Streeter & Phelps, 1925). They came into widespread use, in refined form, in the late 1960s and early 1970s (e.g. Hahn, 1971), when the cost of computing declined rapidly. At that time, water quality was primarily based on dissolved oxygen and optimization of waste discharges was directed towards this goal. In the mid-1970s, programmes were introduced in Austria, the FRG and Switzerland to improve the state of flowing waters. In Switzerland and Austria, this is being achieved by in-stream standards, in the FRG by general requirements on treated effluents. In 1988, approximately 70% of the population in Austria, 85% in the FRG and more than 90% in Switzerland were linked to biological treatment plants.

In lakes, the phenomenon of eutrophication was observed at the time of World War I (e.g. Thienemann, 1918) although the role of the minimum factor did not become apparent until later on. In application in the late 1960s and early 1970s, priority was given to lakes, where in most cases phosphorus was or is the limiting nutrient. Examples of counter-measures applied to lakes in Central Europe were wastewater diversion in Austria and simultaneous precipitation of phosphorus in wastewater treatment in Switzerland.

Estuaries and inland seas came into the focus of the sanitary engineering profession only locally (e.g. San Francisco Bay and the Japanese Inland Sea) and not earlier than the 1970s. The algal blooms observed in 1988 in the eastern part of the North Sea and the northern Adriatic, however, have been the cause of a strong public movement for the protection of inland seas. There are investigations in Denmark (Harremoës, P., 1987, pers. comm.) which show that nitrogen is the limiting nutrient for the seas surrounding Denmark. Thus, despite the fact that we do not have a complete cause–effect relationship for these events, substantial reductions in nitrogen and phosphorus reaching the seas have been proposed by environmentalists and have been taken up by policy-makers.

The compartment 'land/soil' is a less obvious one in relation to water pollution control. However, most of the matter that enters wastewater has its origins there, and as matter can only be converted, there has to be an outlet from wastewater treatment to this compartment. For most water-related sanitary engineers this compartment is an unknown one. Out of 142 million km^2 of land surface on this globe, some 14 million km^2 are utilized for crop production, and the net primary production is estimated to be 9×10^9 Mg of organic matter per year (Lieth & Whittaker, 1975). If all people ($6{\cdot}0 \times 10^9$) had wastewater treatment—including industrial production with a sufficient amount of pretreatment or present-day production technology—less than 1% of the net primary production would be the amount of organic dry substance in residues from biological wastewater treatment. This is a comparatively low figure; local discrepancies are caused by the imbalance between agricultural land and the growth and size of urban centres. In Central Europe, a strong movement for soil protection came into being in the 1970s.

FLOWS OF MATTER AND FEEDBACKS IN THE SYSTEM

Any feedback requires that we start from final recipients. The first flows to be considered start from formation of wastewater and wastewater treatment to land/soil, and the controlling loops feed back along the same paths. The residue from wastewater treatment is waste sludge, which contains valuable materials (calcium, nitrogen, phosphorus, other inorganics, but also organics) and also some matter which can be classified as hazardous (pathogens, heavy metals, toxic organic refractory compounds). This matter, which can be classified as either being directly hazardous or becoming hazardous by transformations in sewage or biological treatment, enters wastewater with its formation. It accumulates in waste sludge in treatment either as solids (it can already be in solid form from the beginning or it is transformed into solids in sewage or wastewater treatment) or, if dissolved, it can be sorbed to floc surfaces of the sludge (e.g. certain heavy metals and many refractory organic compounds).

Whether these toxicants actually behave as such in the soil and the life it supports also depends on the concentration in the waste sludge. Present regulations permit roughly 0·5% of the dry substance of waste sludge to be toxicants, and this limit is getting lower over time. Nevertheless, in public opinion in Central Europe, waste sludge is a mixture of toxic substances with some valuable ones. The newest trend in public opinion-making is to classify waste sludge as a hazardous waste *per se* and to ban it completely from application to soil. Similar tendencies make it more difficult to find controlled landfills for truly hazardous waste sludges, with the consequence that the only possible route becomes incineration and disposal of ashes. However, this route simply shifts the protection of land/soil from this environmental segment to other ones, in an indirect fashion by erection, maintenance and operation of the infrastructure necessary. If the cost for running the infrastructure for land application is set to 1, incineration and disposal of ashes comes out between 5 and 10, and in specific cases even more. For comparison, biological treatment (including removal of biological carbon, denitrification and simultaneous precipitation of phosphorus) holds a value on the same scale of around 6 and the Austrian gross national product is around 4000. The appropriate route for solving this dilemma is to limit the input of hazardous substances into wastewater (link to 'formation of wastewater'), to establish permissible levels of toxicants and to actually monitor this link. Such an approach discards environmental technology that competes with ecosystems and gives employment to smaller institutions.

The next important flows are those from sewage, wastewater treatment and land/soil to the aquatic environment, and the feedbacks along the appropriate paths. As already mentioned, the approach taken in Central Europe with flowing waters, which for most treatment plants are the respective recipients, either starts from in-stream standards or from general requirements for treated effluents. In-stream standards, however, do not protect recipients in the far distance (e.g. estuaries or inland seas) if the values permitted in-stream are too high. For example, the River Danube at Vienna has yearly average values of 2·34 mg/litre for NO_3^-–N, of 0·26 mg/litre for (NH_3/NH_4^+)–N and of 0·02 mg/litre for NO_2^-–N for the period 1982–6, whereas the respective figures in the Austrian in-stream standards are 8·0 mg/litre, 0·5 mg/litre and 0·05 mg/litre. This situation indicates that either the Austrian standard for nitrate is far too high for flowing waters, or we simply have to take into account in our policy-making the allopathic effect of nitrates.

A recent trial to differentiate the origins of total nitrogen (totN) and total phosphorus (totP) in the River Danube Basin at the gauge Vienna-Nußdorf (Fleckseder, 1989) shows that roughly one-third of totN and two-thirds of totP are wastewater-related whereas the remainder is from areal contributions and, with respect to totN, from air pollution. This observation, which is by no means new, tells us that the flow of matter from land/soil to the aquatic environment is genuine and, in combination with the events in the North Sea and the northern Adriatic, that the appropriate control loops must be implemented now.

Finally, consider the link from formation of wastewater to the aquatic environment. Of importance in this link and its associated feedback are those wastewater constituents which enter wastewater in its formation, pass through wastewater treatment and have—directly or indirectly—toxic effects in the aquatic environment. These wastewater constituents are primarily refractory organic compounds. The important task is to stop the transfer of these constituents into the wastewater in the first place. In respect to production, this task is much easier than the wise use of marketed products. The present approach in Central Europe favours legislation and administrative monitoring of the established rules.

AN EXAMPLE CASE: NITROGEN

Figure 1 can also be used as a guideline for the flow of any type of matter that interests us. In this example the transfer and transformation of nitrogen is traced from the formation of wastewater to the recipients.

As mentioned in the description of the compartment 'aquatic environment', we have to consider the co-eutrophicational effects of nitrogen in inland seas. In the past, large international scientific meetings have been devoted to nitrogen (e.g. Nitrogen as a Water Pollutant, 1977) and now the time has come for action. The situation described for the gauge Vienna-Nußdorf on the River Danube is not the same in every catchment to be considered. At this gauge, some 12·5 million people live in the catchment area of 101 700 km^2, and of this area, as an order of magnitude, more than 50% is under agricultural production (tillage, pasture, etc.). As stated above, two-thirds of totN—predominantly in the form of nitrate—is area-related, and one-third is wastewater-related (Fleckseder, 1989). The sanitary engineering profession must react to these figures.

In sewage, nitrogen originates primarily from human metabolism. Other sources are from certain industrial activities, such as production of chemicals, coking, metal surface treatment, rendering and so on, but these sources normally do not enter sewage. Thus, in wastewater treatment, the present applied task is not only to convert ammonium/ammonia to nitrate in order to reduce the toxic risks of ammonia and nitrite, but also to reduce nitrate to gaseous nitrogen. The technology to do this exists, and was developed here in Austria,

TABLE 1
Ratios of COD/totN for Various Wastes or Wastewaters

	$COD_b/totN$ (approx.)
Excreted from humans	3
Wastewater from households	6
Sewage, i.e. the common wastewater from towns	10
Requirements for heterotrophic growth	75
Effluents from sulphite pulping	2000

among other places in the world, in the form of single-sludge nitrifying/denitrifying activated sludge plants. The largest sludge plant in the world to date which operates according to the process and type of equipment which Matsché studied is the plant for the Greater Tel Aviv Area (Gruber *et al.*, 1988). The purpose there was not only directed towards environmental protection, but towards water re-use for potable supply due to elevated nitrate levels in the groundwater.

However, efficient simultaneous nitrification–denitrification is not only a question of technology, it is also a question of the ratio of biodegradable COD (COD_b) to totN in the influent to a treatment plant as well as a question of return flows from sludge treatment and sludge handling in a plant. This ratio can present difficulties. Table 1 shows the ratios of COD_b/totN for the respective types of wastes or wastewaters before they enter treatment plants. In order to have sufficient potential for denitrification, this ratio should be, as influent to an activated sludge unit, not less than 10. These figures indicate, therefore, that sewage is amenable to nitrification–denitrification. However, it is common practice to have primary treatment (sedimentation) in the treatment plants, which reduces COD to a larger extent than totN. In addition, requirements are becoming more and more stringent in sludge treatment (e.g. pasteurization or other types of heat treatment) which in turn means increasing return flows relatively rich in nitrogen over COD_b. In addition, the Western human diet is still high in proteins (compared to carbohydrates), and more and more ready-made meals are entering the market, thus increasing totN and reducing COD_b originating from households.

This present-day situation would be completely different if human metabolic end-products were not included in wastewater, but were collected as undiluted as possible and disposed of in a true ecological fashion on land, as we do with manure from animals. (This would not only be advantageous in terms of nitrogen, but would also help the phosphorus equation as well as reducing the use of water.)

As a sanitary engineer, I am aware that this proposal can, at this point in time, be implemented only in very specific cases and does not help now in the majority of applications. In addition, this idea questions several fundamental principles of the hygienic and sanitary engineering professions which have been unquestioned for several generations, although they have been questioned by environmentalists in the last decade.

This example shows the importance of reconsidering the questions around formation of wastewater and the applied consequences this can have.

CONCLUSIONS

In introducing this paper, the position of wastewater treatment in the context of water pollution control was questioned.

The outcome is that:

(a) wastewater treatment, understood as an end-of-pipe method of obtaining a cleaner effluent and residues from treatment, will hold its importance in the future;
(b) in addition to wastewater treatment, the whole 'complex' of formation of wastewater should be reconsidered in all its consequences to the system presented;
(c) it is important to consider not only the link from wastewater treatment (or from sewage) to the aquatic environment, but also the link to land/soil;
(d) in water pollution control the link from land/soil to the aquatic environment is of prime importance.

REFERENCES

Aldern, E. & Lockett, W. T. (1914). Experiments on the oxidation of sewage without the aid of filters. *J. Soc. Chem. Ind.*, **10,** 523–39.

Barnard, J. L. (1975). Biological nutrient removal without the addition of chemicals. *Water Research,* **9,** 485–90.

Dunbar, W. Ph. (1907). *Leitfaden für die Abwasserreinigungsfrage.* R. Oldenbourg-Verlag, München and Berlin.

Emde, W. v. d. (1984). Biologische Abwassereinigung—Wege und Irrwege, Träume und Wirklichkeit. Lecture held at TU Hannover, FRG, 20 July, in connection with the awarding of the degree 'Dr.-Ing. h.c.'.

Fleckseder, H. (1989). Tracing the sources of nitrogen and phosphorous in the catchment area of River Danube at Vienna, Austria. In *River Basin Management V,* ed. H. Laikari. Pergamon Press, Oxford, pp. 229–36.

Gruber, Y., Farchill, D., Arueste, G. & Goldstein, M. (1988). Start-up and running-in of the Soreq biological wastewater treatment plant. *Water Sci. Tech.,* **20**(4/5), 63–75.

Hahn, H. H. (1971). Mathematische Modelle und Methoden in der Wassergütewirtschaft. In *Wasser-Kalender* 1971, ed. H. Hübner. Erich Schmidt Verlag, Berlin, pp. 18–41.

Lieth, H. & Whittaker, R. H. (1975). *Primary Productivity in the Biosphere.* Springer-Verlag, Berlin.

Matsché, N. F. (1972). The elimination of nitrogen in the treatment plant Vienna-Blumental. *Water Research,* **6,** 485–6.

Nitrogen as a Water Pollutant (1977). Proceedings of the Conference held at Copenhagen, 18–20 August 1975. *Prog. Wat. Tech.,* **8**(4/5).

Streeter, H. W. & Phelps, E. B. (1925). *A Study of the Pollution and Natural Purification of the Ohio River.* US Public Health Bulletin No. 146. US Public Health Service, Washington, DC.

Thienemann, A. (1918). Untersuchungen über den Sauerstoffgehalt des Wassers und die Zusammensetzung der Fauna in den norddeutschen Seen. 1. Mitteilung. *Arch. Hydrobiol.,* **12,** 1–65.

Thomas, E. A. (1962). Verfahren zur Entfernung von Phosphaten aus Abwässern. Schweiz. Patent No. 361.543, 15 April.

10

New Trends in Clean Technologies: Non-Polluting Recycling Technologies. Legislative and Institutional Initiatives in the USA

MARGARET M. KELLY

Director, Technology Staff, Office of Program Management and Technology, Office of Solid Waste and Emergency Response, The United States Environmental Protection Agency, 401 M Street, SW, Washington, DC 20460, USA

Between the years 1984 and 1986, major legislative changes occurred in the United States that had a profound influence on the nation's approach to the management of hazardous waste and non-hazardous municipal solid waste. One such legislative initiative was the passage of the Superfund Amendments and Reauthorization Act (SARA), which revised our Superfund Program to create a strong preference for innovative treatment methods that permanently detoxify contaminated soils. These new methods for waste treatment are mandated to come not only from Federal research and development programs but also from joint ventures or partnerships between Federal and State governments and private industry.

An example of a new cooperative technology transfer program that is assisting in this effort is the US Environmental Protection Agency's (EPA) 'Superfund Innovative Treatment Evaluation' (SITE) program. The SITE program is implementing SARA requirements through demonstrations of innovative technologies at various Superfund sites. A number of alternative treatment methods have already been applied to actual clean-up situations, and we are developing a reservoir of treatment methods and information that will have applicability at sites where clean-up procedures are being planned.

Related to legislative initiatives is a major shift in national policy that changes our waste management approach from 'remedial' to 'prevention'. Three factors are primarily responsible for this change:

(1) *Capacity short-falls are anticipated for non-hazardous wastes.*

Approximately 40% of municipal solid waste landfills in the US will reach their projected capacities by 1995. Seven per cent of the total identified landfills in the US are filled to capacity now. Another 32% have a useful life of 1–6 years (Council on Plastics and Packaging, 1989). Lack of suitable land and growing public opposition to establishing new landfills has created a perspective of urgency for many States where population density is greatest, particularly in the north-east United States.

(2) *Fear of liability for damage to human health and the environment stimulates industry to reduce generation of hazardous waste.* In 1984 the accidental release of 'methyl isocyanate' from a chemical facility at Bhopal, India, created a cloud of toxic vapors that left more than 2000 people in the surrounding community dead, and several thousand injured. In the US, the concern about such accidents intensified when, not long after, a large release of the chemical 'aldicarb oxime' occurred at Institute, West Virginia fortunately without loss of life.

A major financial settlement has been recently required in the Bophal accident—restitution totalled hundreds of millions of dollars.

Although not yet pronounced in cases before US courts, there will be, in speculative likelihood, decisions requiring those responsible for Superfund sites to pay large sums to individuals exposed to toxic contaminants through ingestion of drinking water or the air they breathe.

The potential for financial liability of corporations or companies responsible for mismanaging their waste products is a very real argument and incentive in today's business world for reduced generation of hazardous wastes.

(3) *High costs for site clean-up and treatment of hazardous waste stimulate reductions in waste generation.* An average Superfund site clean-up can cost tens of millions of dollars. On occasion, such costs have risen to 100 million dollars. The treatment of routinely generated hazardous waste by operating facilities has increased in cost because recent land disposal restrictions require a high level of treatment prior to the placement of waste in a regulated landfill.

There is also growing recognition on the part of the public that

Americans in general are consuming natural resources wastefully. Statistics show that in 1960, Americans generated waste at a rate of 2·65 lb per person per day; by 1986, that figure had jumped to 3·58 lb. (US EPA, 1989, p. 12). This trend is projected to continue into the year 2000 unless we take affirmative action to bring about change in this practice. Generation of every kind of waste has risen—paper, plastics, glass and metals. Familiar case studies of industries, such as the electroplating industry, told us that waste of valuable metals was a serious problem.

To combat this trend, EPA has set national goals to address the problems of hazardous waste generation, the growth in municipal solid waste, and specific problems such as chemical releases. Implementation of the goals has resulted in a new commitment at EPA—an incentive-based approach to pollution prevention.

EPA has ranked its preferences for waste management in the following order:

(1) *Source reduction and closed loop recycling.* These are clearly the preferred options. Waste is reduced or eliminated at the source, usually within a process. Specific measures can include process modifications, feedstock substitutions, improvements in feedstock purity, housekeeping and management practice changes, increases in the efficiency of equipment, and recycling within a process. Source reduction is not always 'high technology'. The objective is not necessarily to reach 'zero pollution' but to achieve a net reduction in waste emissions.

(2) *Environmentally sound recycling.* This effort involves the use or re-use of waste as an effective substitute for a commercial product or as an ingredient or feedstock in an industrial process. This includes the reclamation of useful constituent fractions within a waste material or the removal of contaminants from a waste to allow it to be re-used.

(3) *Treatment.* Treatment is defined as any method, technique or process that changes the physical, chemical or biological character of any waste so as to neutralize such waste, to recover energy or material resources from the waste, or the rendering of such waste non-hazardous, less hazardous, safer to manage, amenable for recovery, amenable for storage, or reduced in volume.

(4) *Disposal.* This is the discharge, deposit, injection, dumping, spilling, leaking or placing of waste into land or water so that such waste or any constituents thereof may enter the air or be discharged into any waters, including groundwater.

EPA's approach to promote waste minimization for hazardous wastes at this time is to exploit environmental incentives first and possibly regulate later. EPA has concluded that a mandatory program for waste minimization is not desirable because:

(a) it would second-guess industry's production decisions, possibly leading to counterproductive results;
(b) mandatory programs are difficult and expensive to design and administer; and
(c) generators already face strong economic incentives to reduce their wastes.

A regulatory program would take time to develop, and many industries might postpone any action until mandatory requirements were spelled out. At EPA, we believe the time for making constructive source reduction and recycling decisions is now, while industry is making long-term decisions concerning the land disposal restrictions program (US EPA, 1987).

In general, EPA is taking a multimedia approach to this problem, promoting 'pollution prevention' not just in its waste management programs, but in programs designed to protect surface waters, clean air, and control the use of chemical products such as pesticides. To resolve many of the problems we have encountered in promoting a full program of pollution prevention it is essential that we have not only the cooperation of industry, but also the ability to share resources and information. We hope to create incentives for pollution prevention and eliminate barriers to pollution prevention in existing and planned Agency regulations. The shaping of current and future regulations and identifying research needs to support pollution prevention efforts are critical initiatives.

Initially, EPA is focusing on three themes:

(1) *Education and outreach* programs will include a national clearing house for waste reduction and recycling information. One element is a centralized nationwide repository of waste reduction information from a variety of sources, including all

branches of government, private industry, non-governmental public organizations, and the full range of international waste minimization programs. Case study descriptions are being developed that will document the industry, feedstock, contact at the facility, technologies, affected wastes, environmental medium involved, and costs and savings to the companies involved (US EPA, 1988*a*).

EPA is supporting the United Nations Environment Programme (UNEP) in its effort to create a worldwide network for exchanging information on 'clean' technologies. We will help develop the international database by providing advice on strategies for collecting appropriate waste minimization information. We are also abstracting industrial monographs from the UNEP compendium on low- and non-waste technology, and will make them available through the US clearing house.

Other components of the education and outreach theme are formation of an industrial strategy advisory group; a hotline and newsletter; pro–active communication of concepts to industry at all managerial levels; recognition of accomplishments through an awards program; development of focused curricula for engineering schools; and, working with Congress, industry associations, and public interest groups to foster the national agenda.

(2) EPA supports the initiation and *development of State and local multimedia pollution prevention programs,* encouraging and strengthening the States' ability to work directly with companies and local communities to prevent pollution. Direct financial grants to States for technical assistance programs, combined with training, are an effective way to make States self-sufficient. Creating networks of technical experts and regional demonstration programs of successful technologies are also effective in transferring technology information.

(3) An essential element of pollution prevention programs is the *evaluation of progress* in reducing volumes of waste. EPA will develop reliable 'indicators of waste reduction' and implement a data collection and evaluation strategy to measure progress. We also intend to examine existing databases to determine where information gaps exist, and develop a strategy to fill those gaps. Measures will be developed to evaluate the effectiveness of EPA efforts.

> EPA recently completed a survey of approximately 10 000 hazardous waste generators regarding the volume and characteristics of hazardous wastes produced, and the use of recycling and recovery practices. The information collected will be used to understand each facility's waste reduction program, to determine the success of the program in reducing waste volume and toxicity, and to determine its effect on different environmental media. It will also help EPA establish baseline estimates of waste generation rates and trends.

To implement this new policy, a Pollution Prevention Office has been formed, headed by senior agency managers, to work within the EPA. Gerald F. Kotas is the Director of this office, and has established an executive level Pollution Prevention Advisory Committee to help direct the office and help set priorities in its coordinating and outreach roles.

An additional stimulus for waste reduction is the Chemical Emergency Preparedness Program at EPA, that carries out planning to prevent chemical accidents such as the one that occurred at Bhopal. By legislative mandate (US Govt, 1986), EPA is compiling a Toxic Release Inventory, based on industry-submitted data. The Inventory collects information about minimization activities on a chemical-specific basis. Manufacturing facilities with ten or more employees must report annually on environmental releases of more than 300 listed toxic chemicals. Presently, EPA has information on more than 20 000 facilities.

One innovative response to the program is the realization by manufacturers that there is a market in providing services to downstream users, such as audits directed at chemical safety.

In January 1988, EPA set a national goal for non-hazardous solid waste of 25% source reduction and recycling by 1992. The remainder will be handled by combustion and landfills. Combustors are projected to handle about 20% of the waste stream. Currently only about 10% of solid waste is recycled and 10% combusted (US EPA, 1989).

In keeping with our role as a catalyst and coordinator, EPA has worked with the Environmental Defense Fund, a public interest organization, to generate a national advertising campaign on recycling. We are promoting the procurement of recycled goods, marketing development studies for numerous commodities and promoting non-profit regional market information councils as well as better separation and collection of plastics and lead acid batteries. Some local governments have already imposed waste management 'user charges' levied on households and businesses, based on the garbage generated—the 'pay as you throw' concept.

The final critical component of EPA's approach to enact a policy of pollution prevention is a program of fostering research and development. EPA's Risk Reduction Engineering Laboratory at Cincinnati, Ohio, will be providing industry with information about innovative technologies for reducing hazardous waste. The following OSWER demonstration, evaluation and training endeavors are collaborative efforts between EPA, industry, and other Federal and State Agencies.

(1) *Waste Reduction Innovative Technology Evaluation* has as its primary objective the introduction of waste reduction techniques into commercial practice as quickly as possible. EPA will work with industry to conduct technical and economic evaluation of new and promising waste reduction technologies. To encourage the participation of small to medium-sized companies in this effort, the program provides technical and financial support through State and local government agencies.

(2) *Waste Reduction Evaluations at Federal Sites* focus on waste minimization. These consist of a series of demonstration and evaluation projects for waste reduction conducted cooperatively by EPA and the US Department of Defense, the US Department of Energy, and other Federal Agencies. Results at these sites can be transferred to the private sector.

(3) *The Waste Reduction Assessments Program* encourages industry to use waste minimization assessments as a tool for identifying reduction opportunities. To assist industry in conducting these assessments, EPA's Office of Research and Development has developed a *Manual for Waste Minimization Opportunity Assessments*. Applicable to various industrial settings, the manual provides a uniform basis for industries to implement, postpone or reject identified waste minimization options through economic analysis. As more industries use the manual, EPA will evaluate the effectiveness of translating incentives and disincentives for waste minimization into quantifiable factors.

The State of New Jersey is currently field-testing the manual at 30 facilities. EPA plans to tailor the manual to the needs of specific industries.

(4) A panel of experts in the principles and practices of waste minimization will comprise the *Waste Reduction Institute for Senior Executives*. Sponsored jointly by EPA and various US universities, panel members will encourage private industry generators to use reduction techniques.

An additional important element of EPA's strategy is the training of State inspectors and industry representatives in waste minimization auditing. The objective of the training is to develop in-house capability at a manufacturing or processing facility to identify opportunities for waste minimization. The program will offer workshops targeted to specific industries across the nation this year. The workshops will apply a standard set of audit procedures, given in the EPA *Manual for Waste Minimization Opportunity Assessments,* to industrial situations. Specific sessions will address actual cases outlined earlier by participants, and hypothetical cases developed by instructors.

In addition, the workshops will provide instruction in conducting cost/benefit analyses, addressing management concerns, identifying incentives for waste minimization within a specific facility, and identifying information sources within EPA that illustrate innovative minimization strategies. The industries addressed by the workshops will represent the fields of electronics, petroleum refining, plastics, chemicals, metal finishing and fabrication, wood preserving and textiles (US EPA, 1988*a*).

Industry's tradiational approach to pollution abatement has been to utilize costly and energy-intensive add-on control devices. In the last 5 years, however, a number of companies report innovative changes. These tend to be the largest companies. A number of facilities are making process changes and changing patterns of storage of inventories to avoid the risk of a chemical accident and to avoid, whenever possible, additional legislation and subsequent regulation.

One example of a waste minimization program that has had measurable success comes from the 3M Corporation's 'Pollution Prevention Pays' (3P) program. 3P is an innovative program begun in 1975 to eliminate pollution at the 'source-in' end of the product development and manufacturing process. It is by now relatively well-known and incorporates a number of characteristics that serve as a model for successful waste minimization programs:

(1) 3P is an employee incentive program which gives 3M Corporation technical employees the opportunity to provide innovative ideas that make pollution prevention a company policy. Employees apply their ideas to product formulation, modification of manufacturing processes, redesign of equipment, and recycling of materials used in processes.
(2) The 3P program is judged to be highly successful by quantitative measures. 3M Corporation estimates that the program has resulted in the following annual emission reductions at its facilities throughout the US: 110 000 tons (US) of air pollutants; 13 000 tons (US) of water pollutants; 275 000 tons (US) of sludge and solid waste; and 1·5 billion gallons (US) of wastewater.
(3) The total worldwide financial savings, to date, exceeds 420 million dollars (estimated for the first year of the project's operation). These savings include pollution control equipment that did not have to be purchased; materials and energy saved; and sales from reformulated products.
(4) 3M Corporation proposed projects are reviewed quarterly by a five-member coordinating committee. Projects must have management support evidenced by a commitment of funds. A project must be innovative or incorporate a unique or original design, or involve a significant technical achievement (US EPA, 1988*b*).

One final example of innovative private sector response to non-regulatory incentives is waste exchanges. There are a number of information exchanges in different regions in the United States. One of these, the 'Northeast Industrial Waste Exchange' has been in operation since 1981 (Northeast Industrial Waste Exchange, 1988). This exchange matches waste generators with waste users. The goal of the exchange is to recycle waste materials back into the manufacturing process. Participation can minimize waste disposal expenses; reduce the need for landfilling; and increase the volume of wastes, by-products, and surplus and off-spec materials.

The Exchange publishes a quarterly catalog in which wanted and available materials are listed as coded, classified advertisements. The catalog is distributed throughout the United States and Canada, with a circulation of more than 14 000 subscribers. The catalog is also 'computerized' for up-to-the-minute information. A customized list of recyclers who handle waste materials is also available. The recyclers will meet specifications for geographic range, packaging and materials contamination.

The US goals for waste minimization, or more broadly, 'pollution prevention', are based on a partnership between industry and Federal and State governments. The stimulus is not regulation, but an awareness that pollution prevention is the only route that makes sense in a world where our oceans, rivers, groundwater and air are unable to absorb additional environmental insults.

BIBLIOGRAPHY

Backman, Huisingh, & Siljebratt (1988). Preventive environmental protection strategy: Preliminary results of an experiment in Landskrona, Sweden. TEM/University of Lund.

Council on Plastics and Packaging (1989). *Environment Quarterly*, **3** (Winter), 1.

Northeast Industrial Waste Exchange (1988). *Listings Catalog,* Issue No. 30, November.

US EPA (1987). Waste minimization: Environmental quality with economic benefits, October, p. 6.

US EPA (1988*a*) Pollution prevention review. Winter.

US EPA (1988*b*) Cooperative Environmental Management Message.

US EPA (1988*c*). Review of emergency systems: Report to congress, June.

US EPA (1988*d*). Summary of survey responses on changes in technologies and practices, technologies under development, and technologies in need of development—Discussion paper.

US EPA (1988*e*). Waste minimization opportunity assessment manual, July.

US EPA (1989). The solid waste dilemma: An agenda for action, February.

US Govt (1986). Emergency Planning and Community Right-to-know Act, Section 305(b), Superfund Amendments and Reauthorization Act.

11

Strategy for Hazardous Waste Management

MASARU TANAKA

The Institute of Public Health, Ministry of Health and Welfare, 6-1 Shiroganedai 4-chome, Minato-ku, Tokyo 108, Japan

CLASSIFICATION OF WASTES

In the Waste Disposal and Public Cleaning Law of Japan, enacted in 1970, wastes are classified as shown in Fig. 1. (Radioactive wastes are handled separately.) According to this law, wastes are classified into general wastes from living activities and wastes from business activities. In addition, many terms such as hazardous (industrial) waste, medical waste, etc. are used. It should be remembered that the definition of these terms is quite different in different countries and in Japan there is often a gap between a legal term and the commonly-used term.

This classification is important as it relates to disposal responsibility, burden of costs, etc. In the case of general waste, the local authority (city, town and village) is responsible for waste disposal planning, arrangement of disposal facility and its management. However, industrial waste must be disposed of by the waste-generating business itself. (Needless to say, the consignment to a management contractor, who is engaged in the disposal, is admitted.) With regard to costs of disposal, the cost for general domestic waste is paid by the ordinary account of the local authority, but the cost for wastes from business activities is usually paid by the waste-generating business operator.

Commercial values of merchandise vary considerably. Some items may be transacted at several thousand yen per ton, while some other items may be transacted at several tens of thousand yen per gram.

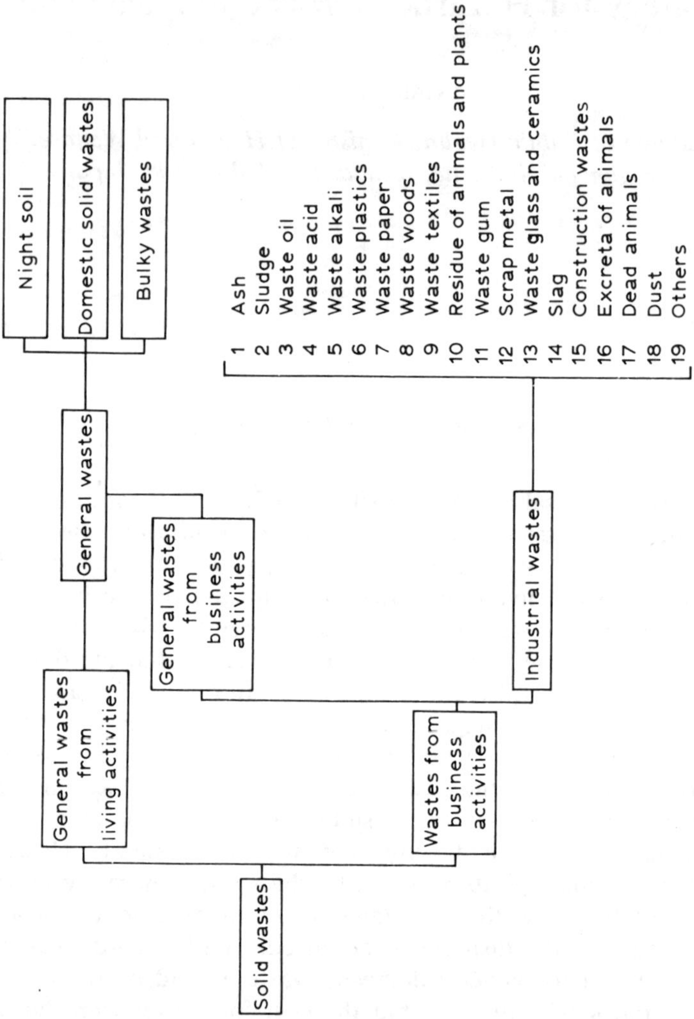

FIG. 1. Classification of solid waste (Waste Disposal and Public Cleaning Law).

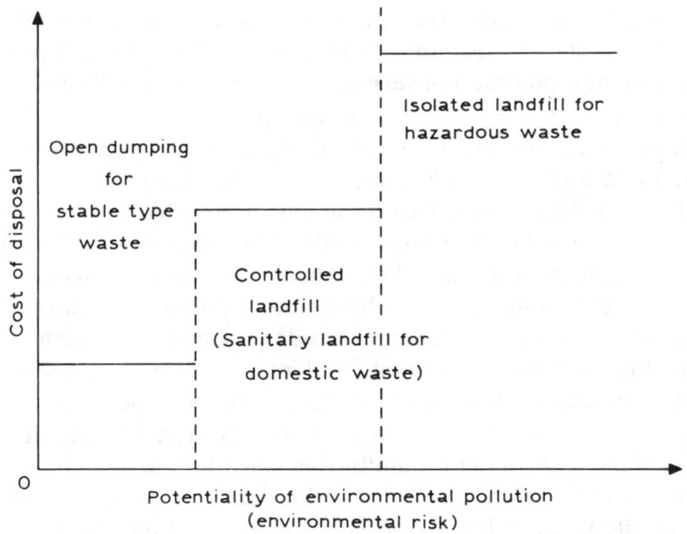

FIG. 2. Classification of waste based on environmental risk.

Similarly, when the cost of waste is measured with the adverse effect to the environment, the value varies depending on its content.

In Fig. 2 the possibility and the degree of environmental pollution generated is called 'potentiality of environmental pollution' or 'environmental risk'. It is classified into three categories according to the type of landfill required. The required landfill structure for each type is different.

The most risky waste is called 'hazardous industrial waste'. It contains heavy metals and other materials which may damage people's health, and the risk of leaching from hazardous industrial waste is high. If these items are disposed of without any pretreatment, they must be landfilled in an isolated landfill site. The facility must be made of concrete and have an undamaged structure. It must also be isolated from the surrounding environment and have a waterproof structure, thus eliminating pollution by leaching.

On the other hand, wastes which can be disposed of at stable type landfill sites are: waste plastics, waste gum, scrap metal, waste glass, waste ceramics and construction wastes which are chemically and biologically stable and do not generate polluted leachate. The degree of environmental pollution is very low. This disposal facility is called a 'stable-type landfill site' or an 'open-dumping facility'. Since leaching is not a problem, the facility does not need to be waterproofed and collection and treatment of leachate are not required.

As shown in Fig. 2, non-hazardous but non-stable-type waste must be disposed at a controlled landfill site. The leachate will pollute the environment unless it is controlled. Therefore, leaching shielding, and collection and treatment of leachate are required. The waste to be disposed of in this type of landfill includes waste paper, waste woods, waste textile, residues of animals and plants, animals excreta, dead animals, non-hazardous cinder, sludge, slag, etc. General waste or municipal waste must be disposed of in this type of landfill. Some countries require this type of landfill for hazardous waste.

The horizontal axis in Fig. 2 shows the degree of environmental risk caused by the waste if left in the environment without any treatment. When the required landfill disposal is implemented, the environmental risk caused by the disposal facility is not necessarily in the same sequence of wastes, i.e. wastes for isolated landfill, wastes for controlled landfill and wastes for stable-type landfill. The environmental risk from landfill sites appears to be in the reverse sequence. In the isolated landfill site, because the waste is separated from the environment with concrete, under normal conditions, the possibility of environmental pollution is zero. For the controlled landfill site, seepage is controlled and wastewater is collected and treated. Then, the treated water is discharged to the water body. If environmental damage occurs, it can be assumed that the seepage control work does not function or the treatment of leachate is not sufficient.

In the case of the stable-type landfill site, the problem is that of quality control (QC) of the waste. It is a feature of these wastes that they are mixed and the possibility that hazardous chemical material or organic material is included cannot be denied. It cannot be guaranteed that the leachate from a stable-type landfill site will not pollute the environment. If the waste plastics, scrap metal, waste glass, etc. do not include the above mentioned materials, and wastewater is not generated when they are soaked in water, they may be recycled and may not be disposed of as waste. From this viewpoint, the stable-type landfill site is not necessarily the safest one.

WASTE MANAGEMENT AND ENVIRONMENTAL POLLUTION

When we observe the relationship between waste management and environmental pollution, the following factors should be carefully analysed.

Contents of Wastes

In order to know what pollution is brought about, it is important to find out what hazardous material is present and what its leaching potential is. Hazardous materials are considered to be heavy metals such as mercury, etc. and organic chlorinated compounds such as PCBs, etc. Generally as industrial wastes contain considerable quantities of these materials, the confirmation of proper disposal of chemical wastes is essential.

Therefore, the following questions are important in order to find out the potential for environmental pollution:

(1) What types of wastes are discharged by what types of facilities?
(2) What hazardous material is contained in the waste?
(3) What is its leaching potential?

Pollution Media

The processes which bring about damage to the quality of the environment and people's health are considered to come through the media of air, water and soil. Including the material generated during treatment and disposal, it is important to determine what medium the hazardous material is discharged into, what movement is made after discharge, what change is brought to the environmental quality and what level of intake is made to the people.

Mechanism of Pollution

In the flow of waste management, i.e. generation → discharge → collection → transportation → intermediate treatment → final disposal, the possibilies of pollution are as follows:

(1) accident at the stage of collection, transportation, or disposal;
(2) illegal dumping;
(3) excessively low environmental protection level at the disposal stage.

Air pollution from intermediate treatment facilities such as incinerators, water pollution from landfill sites, etc. cause concern. The relationship between waste landfilling and health effects may be explained as follows. Wastes used for landfill generate leachate containing hazardous substances which may be discharged to the water body. These hazardous substances then accumulate in living things and/or contaminate drinking water and finally health effects could be caused.

DEFINITION OF HAZARDOUS INDUSTRIAL WASTES

Hazardous industrial wastes, whose final disposal (landfilling and ocean dumping) are forbidden in the original state except for isolated landfill, are called 'industrial wastes containing metal, etc.' in the Waste Disposal Law. The definition is made according to the type and nature of wastes.

The following eight categories are specified as 'metal, etc.' for the case of landfilling;

(1) mercury and its compounds, alkyl-mercury compounds;
(2) cadmium and its compounds;
(3) lead and its compounds;
(4) organic phosphorus compounds (limited to parathion, methyl-parathion, methyl-demethon and ethyl para-nitro-phenyl);
(5) chromium (VI) compounds;
(6) arsenic and its compounds;
(7) cyanides; and
(8) polychlorinated biphenyls (PCBs).

'Hazardous industrial waste' has to be generated from particular facilities specified according to manufacturing process and raw material used. As a second condition, it must be one of the six specified types of industrial waste. As a third condition, potential leaching of hazardous chemicals (see Table 1). The sludge, waste acid and waste alkali, which are discharged from the following facilities;

(1) Among the specific facilities stipulated in Water Pollution Control Law, the facilities which may discharge wastes containing metal, etc.

(2) Facilities stipulated in "Waste Disposal Law", which may discharge waste containing metal, etc. (other than the above (1)), the cinders, which are discharged from the following facilities:

(3) Incinerator of sludge, waste acid or waste alkali generated in the factory or facilities having specific facility shown in (1) and (2) or incinerator of designated sewage sludge,

(4) Incinerator of waste plastics, etc., which is industrial waste disposal facility, the dust, which are discharged from the following facilities;

(5) Incinerator shown in (3) and (4) above,

(6) Among the specific facilities stipulated in Air Pollution Control Law, the facilities which may discharge waste containing metal, etc.

and all kinds of slag and PCB polluted items could be hazardous waste. Among all, those which contain metal, etc. (that are determined considering raw material, process, etc. of the respective facilities) and which do not suit the 'judgement criteria related with industrial wastes including metal, etc.' shown in Table 1, are classified as 'industrial wastes including metal, etc.' or 'hazardous industrial waste'.

TABLE 1
Standard for Verification of Hazardous Chemical by Leaching Test. (Disposal Method, Landfill; Wastes—Slag, Sludge, Cinder, Dust)

Hazardous substances	*Value (mg/litre)*
Alkyl-mercury compounds	ND
Mercury and its compounds (Hg)	>0·005
Cadmium and its compounds (Cd)	0·3
Lead and its compounds (Pb)	3
Organic phosphorus compounds (O–P)	1
Chromium (VI) compounds (Cr)	1·5
Arsenic and its compounds (As)	1·5
Cyanides (CN)	1
PCBs	0·003
Organic chlorine compounds	—

STRATEGY AGAINST HAZARDOUS INDUSTRIAL WASTES

The purpose of waste management is the improvement of public health, protection of the living environment and conservation of natural resources. When considering the strategy of waste management, the purpose and target of waste management must be clarified. In order to accomplish these targets, any restrictive conditions must be analysed.

In general the following restrictive conditions are considered.

(1) *Land space for final disposal.* As the land area size is small in Japan, the acquisition of sites for landfill is extremely difficult. The level of difficulty depends on whether the area is urban or rural and the distance from the waste generation point.

(2) *Environmental impact.* Waste management has the purpose of environmental protection, but the waste management facility itself has an effect on the environment. Whether this effect obstructs accomplishment of the environmental protection target or not and whether the degree of effect on the environment is acceptable or not must be checked.

(3) *Health impact.* Whether the effect on health of pollutants generated by the waste management facility is acceptable or not must be confirmed.

(4) *Conservation of natural resources.* All over the world, it is understood that the conservation of natural resources is essential. It must be confirmed that the waste management does not form an obstacle for the accomplishment of the higher ranked target, the conservation of natural resources.

(5) *Limitations of equipment, materials and workers.* There is a limit to the number of workers who can be engaged in the waste management, and to the equipment, materials and disposal technology, etc. The accomplishment of the above-mentioned target is thus limited by the availability of equipment, materials and manpower. How effectively the management target is accomplished with limited resources is important.

(6) *Budget limitation.* The budget available for waste management is also limited.

What can be done to accomplish the above-mentioned target with limited resources? Figure 3 illustrates a strategy for hazardous waste

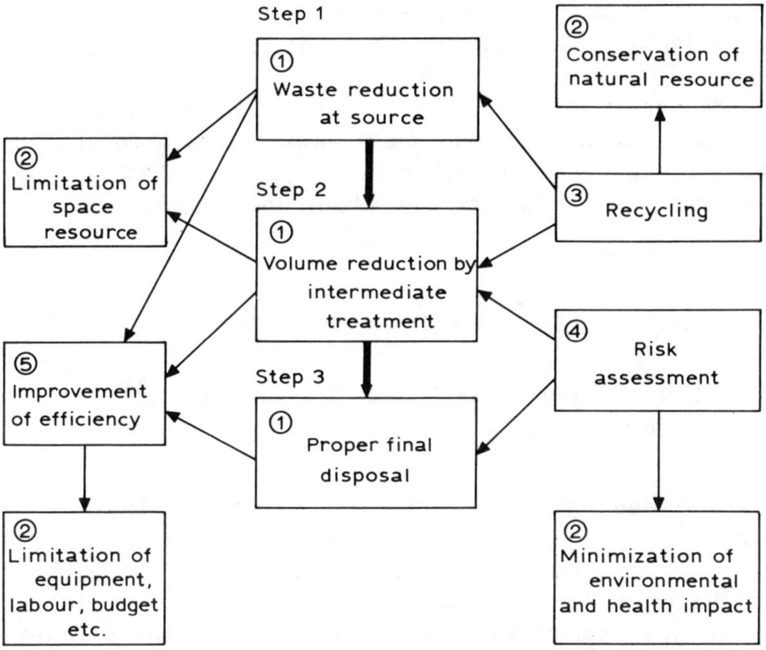

FIG. 3. Strategy for hazardous waste management. ①, Target to achieve; ②, constraints; ③, activity method; ④, evaluating method; ⑤, evaluating indicators. ➞, Next step; →, good effect to.

management. Number ① indicates the targets to be achieved to fulfil the waste management goal. Number ② shows various restrictive conditions for accomplishing the target. Number ③ indicates strategies or tactics necessary to accomplish target ①. The importance placed on recycling activities is evident. By the promotion of recycling, the conservation of natural resources, which is a restrictive condition, can be satisfied to some extent and at the same time, reduction of wastes at the generation source can be accomplished.

Furthermore, a reduction in waste volume can be accomplished by recycling. Number ④ indicates the evaluation method used to determine whether or not the effect on the living environment and health impacts are small enough. With this risk assessment, a scientific overall judgement is made and the proper disposal level will be determined.

In order to accomplish the purposes with the limited resouces (equipment, material and manpower) available and within the restrictions of budget, efficiency must be enhanced. As the evaluation parameter for efficiency, the cost per ton, wastes collected or disposed per worker must always be calculated with relevant parameters and how to enhance the efficiency must always be analysed. The data should be prepared as to whether or not proper and efficient waste management is achieved or how to enhance the efficiency.

SETTING OF DISPOSAL LEVELS

On the setting of appropriate disposal levels of industrial wastes, the following approaches are taken in Japan.

Disposal Technology Corresponding with the Characteristics of Wastes

First, the appropriate technology corresponding with the characteristics of wastes must be selected. From the viewpoint of landfill, there are three types of landfill in Japan as already explained. If hazardous industrial waste is disposed of with no pretreatment, it must be disposed of at an isolated landfill facility. Stable-type waste can be disposed at the stable-type disposal facility, on which the environmental protection technology is simplest. All other wastes are disposed of on the controlled landfill, under which leachate is treated. An appropriate technology, which corresponds with the characteristics of waste (that is, the potentiality of environmental pollution) is adopted.

Acceptable Wastes Corresponding with Disposal Facility

The second approach is based on the fact that the degree of environmental protection varies for the individual landfill facility. The degree of use of underground water is different at different disposal facilities and the degree of environmental protection required by the surrounding citizens is different. Structures, sizes and monitoring system of landfill facilities vary. Therefore, the list of acceptable waste is decided for each disposal facility. Some items are accepted and others are not accepted, in addition to the requirements by law and regulation. In general, the list of acceptable wastes is written in the Pollution Control Agreement.

Environmental Impact Assessment

The third approach is that the effect on the environment is assessed and the minimum level of effect on the environment confirmed. In the case of large disposal facilities, the environmental impact assessment statement must be prepared according to the law. At present, use of a site is not possible, regardless of its size, unless the effect on the environment has been assessed. This assessment of impact on the environment is the so-called EIA. The areas to be assessed are health, living environment and natural environment. For the individual environmental quality, future tendencies are prospected. It has to be confirmed that the background level plus impacts to the environment caused by a specific waste disposal facility during construction, usage or end of usage are lower than the targeted environmental protection level. In fact, the background and the additional impact caused by the facility are added (for example, on pollution criteria such as suspended particular matters, NO_x, SO_x and HCl). When it has been confirmed that no obstacle exists to accomplish the reference level for environmental protection, the construction and operation of the disposal facility can go ahead.

ENVIRONMENTAL PROTECTION REQUIREMENTS AND DESIGN CRITERIA

On the environmental protection at industrial waste disposal facilities the Air Pollution Control Law and/or Water Pollution Control Law are related as shown in Fig. 4. The prefecture authority puts other articles on these laws and local public authority (such as city, town and village) puts more severe articles or stipulates its own articles. Following agreement on pollution prevention with inhabitants around the facilities, the final control conditions on environmental protection are established. In order to secure these conditions, the design criteria of facilities are established considering safety factors, etc.

In the case of landfill sites, the acceptable wastes, number of collection and transportation trucks, quality of discharged water after leachate treatment, etc. will be the content of agreement.

ACKNOWLEDGEMENT

The work described in this paper was supported by the Joint Research Projects of Japan Health Sciences Foundation.

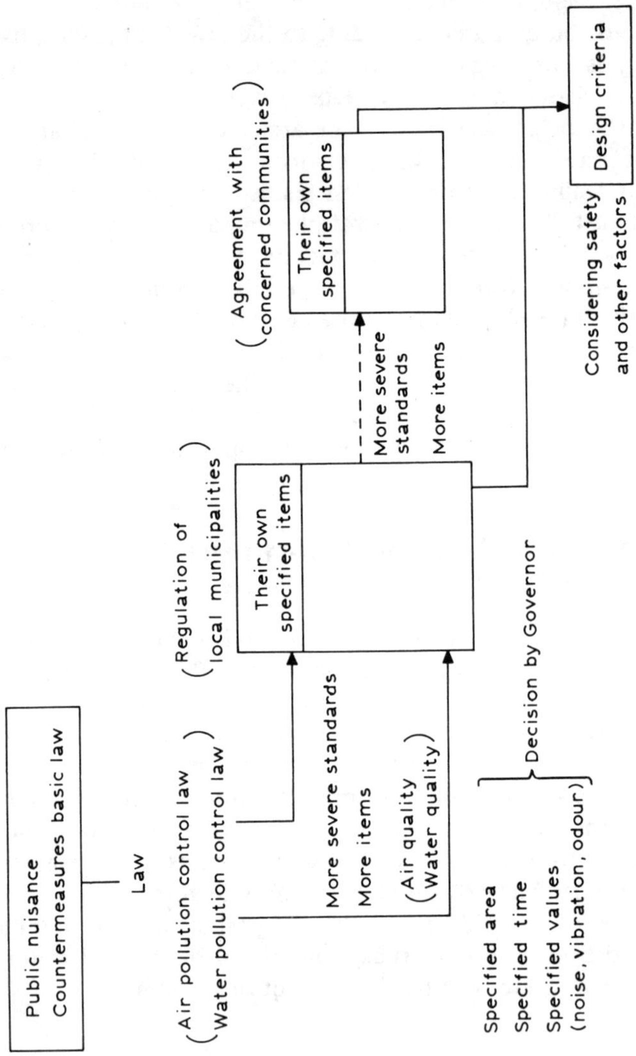

FIG. 4. Relations between environmental protection requirements and design criteria.

BIBLIOGRAPHY

Environment Agency, Japan (1987). Quality of the environment in Japan.

Environment Agency, Japan (1987). Environmental Laws and Regulations in Japan (1) General, (2) Air, (3) Water-I, II, (4) Nature.

Tanaka, M. (1988). Recycling and other alternatives for waste volume reduction. In *ISWA 88 Proceedings of the 5th International Solid Wastes Conference,* Vol. 1. Academic Press, London.

12

Monitoring and Control of Transfrontier Movements of Hazardous Wastes: An International Overview*

HARVEY YAKOWITZ

Consultant to the Environment Directorate, Organisation for Economic Cooperation and Development (OECD), 2 rue André Pascal, F-75775 Paris cedex 16, France

BACKGROUND AND SCOPE OF THE PROBLEM

Wastes are items which are disposed by various means such as burial, insertion into the sea, incineration, spreading onto land, etc. Hazardous wastes are those which, if improperly managed, could harm man and/or the environment because they are toxic, corrosive, explosive, combustible, etc. Often, certain components of wastes, such as heavy metals, cause that waste to be judged potentially hazardous by regulatory authorities. Lists of potentially hazardous wastes have been issued by many countries. In late May 1988, member countries of the Organisation for Economic Cooperation and Development (OECD) agreed upon a list of 44 potentially hazardous wastes. Special monitoring and control is required if any of these wastes cross the frontiers of any OECD member country on their way to final disposal (OECD, 1988).

Monitoring means that the whereabouts of hazardous wastes are known at all times 'from cradle-to-grave' and that the wastes are directed to an appropriate facility for treatment and disposal. Control means that authorities are aware of waste flows and can act rapidly to ensure that the possibility for inappropriate handling of the wastes is minimized.

Millions of tonnes of potentially hazardous wastes cross OECD national frontiers each year on their way to disposal because legal

* The views expressed are solely those of the author and not those of OECD and/or its member countries.

disposal in a foreign country may be less expensive than at home or because there is no local disposal capacity for these wastes. On average, a consignment of hazardous wastes crosses an OECD frontier every 5 min, 24 h per day, 365 days per year. There are over 100 000 such movements in OECD European countries yearly. About 6000 such movements occur annually in North America.

While most of this traffic takes place among OECD member countries, certain quantities of potentially hazardous wastes are exported from the OECD area. In many instances, Eastern European countries are the importers. But in 1988, there were many articles in the press throughout the world reporting the shipment or proposed shipment of large amounts of wastes to countries with developing economies.

There were outcries against such waste traffic. The Organization of African Unity unanimously passed a resolution condemning importatation and disposal of potentially hazardous wastes on African soil. The European Parliament passed a resolution condemning all significant exports of potentially hazardous wastes from the European Community to any country with a developing economy. Legislation was proposed in the United States Congress which would prohibit export of any waste classed as hazardous under US law unless a special permit (export license) was granted stipulating that the waste would be disposed as if it were in the United States.

Why is there such concern and indignation about transfrontier movements of hazardous wastes? One main reason is clear: no nation, indeed no local area, overtly wishes to be the 'garbage dump' for someone else. The second main reason is that the health aspects of exposure to potentially hazardous wastes are, to paraphrase Churchill, an enigma wrapped up in a mystery. This problem is especially important with respect to chronic effects from exposure to wastes which have toxic properties. The uncertainties associated with understanding and combating these effects have created a large 'dread factor' among people.

To date, no definition of adverse health effects from exposure to these wastes has been agreed upon. Overt clinical disease or distinctive syndromic effects are unlikely to result from living close to a site where such wastes have been deposited. Rather, intensification of pre-existing conditions or non-specific deterioration of health and well-being may occur (Phillips & Silbergeld, 1984). Mental stress can lead to disfunction and has reportedly done so in cases of exposure to sites where hazardous wastes have been deposited (Levine, 1982; Young,

1983; Reko, 1984). In most instances, such health aspects investigations must rely on epidemiology as a principal tool. Use of this tool means that adverse health effects, if any can be unambiguously identified, are only revealed *after* damage has occurred. Moreover, epidemiology is expensive, time-consuming and may be difficult to apply, especially in areas where education may be limited.

Added to these difficulties, habits and cultural differences may alter the probability that any individual exposed to a potentially harmful substance will suffer adverse health effects or that such adverse effects, if present, are primarily related to the exposure. These differences include dietary and nutritional aspects, socio-economic status and alcohol and tobacco intake. Hence, exports of wastes which if disposed in a certain fashion in the country where they were generated would be 'safe' may not be 'safe' if disposed the same way elsewhere.

FORCES ACTING TO PROMOTE TRANSFRONTIER MOVEMENTS OF HAZARDOUS WASTES

There are a number of potential stimuli for causing generators of waste to consider export as a means of dealing with these wastes. This list includes, but may not be limited to, the following:

(1) rising costs of disposal in the home country;
(2) diminishing capacity for disposal of certain types of wastes in the home country;
(3) potential future liability for any damages caused by wastes 'stored' in the home country;
(4) tightening of laws, regulations and policies concerning disposal of certain types of wastes, e.g. prescriptive disposal routes, such as incineration being required for liquids containing certain organic constituents;
(5) tightening of laws, regulations and policies governing on-site disposal operations for wastes;
(6) general economic growth which may result in more total generation of wastes;
(7) existence of regional disposal facilities located outside the home country; and
(8) market opportunities for materials which can be recovered, reclaimed or recycled from wastes otherwise destined for 'final' disposal.

For a generator of wastes, the chief consequence of any one or some combination of the items listed above is that legal disposal in the home country will almost certainly become increasingly costly as a function of time. A generator will normally seek least cost legal disposal for his wastes. If export is available, legal and less costly than disposal in the home country, then export is a likely choice. Two key questions arise: (1) What types and quantities of wastes are likely to be prime candidates for export? (2) What cost savings to generators can be expected if export is chosen for these wastes? These two questions were first considered in some detail by the OECD* (see note 1) some years ago (Yakowitz, 1984). In February, 1984, OECD member countries were informed that:

(1) hazardous waste which crosses frontiers and is destined for disposal in another country is likely to be waste considered highly hazardous, i.e. requiring incineration or physico-chemical treatment in the generator country as well as being restricted from legal sea dumping;
(2) within Europe, journeys of up to 800 km can be contemplated by a generator (for hazardous wastes requiring physico-chemical treatment or incineration in the home country) in order to reduce total costs of disposal;
(3) within OECD-Europe, about 2 million t of hazardous waste are estimated to cross national frontiers annually on the way to legal disposal either at sea or ashore. This figure represents 8–10% of all such wastes generated in OECD-Europe;
(4) transport of highly hazardous wastes, e.g. those banned from sea dumping by international conventions, via ship to certain developing countries from industrialized nations, e.g. OECD member countries, could be very profitable in economic terms for all participating parties; hence, the occurrence of such 'north-to-south' movement of highly hazardous wastes should be considered as a real possibility;
(5) implications of possible shipments of highly hazardous wastes to developing countries, i.e. 'north-to-south' movement of such wastes may need to be assessed. Controls by industrialized nations on such exports may need to be studied, e.g. the legality of the imposition of restrictions on highly hazardous wastes destined for disposal in a developing country;
(6) methods providing for monitoring and control of transfrontier movements of hazardous wastes should be developed.

So far as cost differentials were concerned, about 100 European Currency Units (ECU) per tonne was estimated as the early 1984 average cost differential between disposal into or onto land (or insertion into the sea) and incineration and/or physico-chemical treatment of certain classes of hazardous wastes. Today, this figure for liquids and pumpable sludges containing organics and/or heavy metals or cyanidic components is probably more than 200 ECU per t.

In the aggregate, between 2 and 2·5 million t of potentially hazardous wastes probably crossed OECD European frontiers in 1988. If 100 ECU per t is allowed as the cost of paperwork, packaging, labelling, transport and insurance, then marginal cost savings to generators of at least 100 ECU per t of wastes exported can be estimated. Apparently, the aggregate annual marginal 'savings' to generators in OECD Europe represents roughly 200–250 million ECU. The 'trade' perhaps represents a total cash flow of about 500 million ECU.

The distances travelled during transfrontier movements of hazardous wastes average a few hundred kilometres in Europe. The longest recorded distance for a legal transfrontier movement of hazardous wastes is over 17 000 km. (Does not include the voyages of vessels whose cargoes of potentially hazardous wastes were refused the right to unload in a given port of call.)

ACTIONS BEING TAKEN TO MONITOR AND CONTROL TRANSFRONTIER MOVEMENTS OF HAZARDOUS WASTES

OECD

The report of February 1984 to the Waste Management Policy Group of the OECD was only one stimulant for international action by member countries. Indeed, on 1 February 1984 the OECD Council (the governing body of OECD) decided that member countries would have to control transfrontier movements of hazardous wastes. Moreover, a comprehensive set of guiding principles concerning such control was recommended to member country governments (OECD, 1984). These OECD actions provided the foundations for European Community legislation in this area; Directive 84/631 of 8 December 1984 in effect provides for implementation of the principles embodied in the recommendation of the OECD Council.

Following a meeting of OECD Environment Ministers, on 20 June

1985, by means of Resolution C(85)100, the OECD Council decided to:

> develop an international system for effective control of transfrontier movements of hazardous wastes, which will include appropriate OECD instruments such as further Acts of the Council covering notification, identification and control of such transfrontier movements, as well as an international agreement of a legally binding character;

and specifically instructed the Environment Committee (OECD, 1985):

(a) To undertake the work necessary to implement this Resolution with a view to making proposals in the form of appropriate OECD instruments and a draft international agreement before the end of 1987 (see note 2);
(b) To assess, after consultation with other competent international organisations, whether the agreement should be developed in such a way that it can be open to both OECD member countries and other interested countries;
(c) To base the international system upon the principles contained in the Decision and Recommendation of the Council on Transfrontier Movements of Hazardous Waste [C(83)180-(Final)], further developed in the light of the Conclusions and Recommendations adopted by the OECD Conference on International Cooperation Concerning Transfrontier Movements of Hazardous Wastes, subsequently endorsed by the Environment Ministers of OECD Member countries.

With respect to other OECD Council Acts, the Decision–Recommendation C(86)64(Final) of 5 June 1986 addresses the matter of exports of hazardous wastes from the OECD to, *inter alia,* developing countries. It calls upon member countries to prohibit exports of hazardous wastes to non-member countries unless written consent is obtained from appropriate authorities in such countries and unless the wastes will be directed to an adequate disposal facility (OECD, 1986). The Act of the Council (C(88)90(Final) of 27 May 1988) specifies what the terms 'wastes' and 'disposal' mean, specifies what wastes shall be referred to as 'hazardous wastes' and provides for a uniform classification system for these wastes for purposes of the international control system (OECD, 1988).

During its 47th Session on 7–9 December 1988, the Environment

Committee considered a draft International Agreement that essentially draws together these earlier Council Acts and adds the necessary elements (in particular the notification procedures) to create the basis of a comprehensive system to control transfrontier movements of hazardous wastes (OECD, 1989). It embodies the following basic principles:

(a) Those wastes that are referred to as 'hazardous wastes' for purposes of the control system, are clearly identified.
(b) Persons, such as the generator or disposer, who are technically capable of correctly determining if a 'waste' is not 'hazardous waste' subject to the terms of the control system bear the responsibility for such determinations.
(c) Authorities who are competent to judge whether a proposed import of hazardous wastes can be managed in accordance with the laws/regulations and practices of the country of importation must receive prior notification that such an import is foreseen.
(d) Competent authorities of the country of importation must formally confirm that they have no objection to a proposed transfrontier movement of hazardous wastes which is to enter territory under their jurisdiction.
(e) Simplified procedures are possible for hazardous wastes destined for recycling; in particular, if there is no objection, written confirmation is not necessarily required from the country of importation.
(f) All countries through which hazardous wastes are to transit must receive prior notification.
(g) If the authorities of a transit country object to the passage of hazardous wastes, the transfrontier movement cannot legally enter their territory until the objection is lifted.
(h) For transfrontier movements of hazardous wastes, only transporters and disposers who are authorized or allowed to perform such operations may be employed.
(i) If a transfrontier movement of hazardous wastes cannot be completed as foreseen, efforts to ensure that the wastes are disposed in an environmentally sound manner must be undertaken. In particular, contracting Party countries of exportation may not hinder or oppose return of the wastes.
(j) Control of transfrontier movements of hazardous wastes to countries which are not Contracting Parties should be no less

protective of the environment than control of such movements among Contracting Parties.

(k) Bilateral or multilateral agreements or arrangements are permitted provided that any such agreements or arrangements are compatible with the principles of the International Agreement.

(l) Provisions exist to provide for regular, periodic information exchange concerning all aspects of implementation and functioning of the International Agreement.

The proposed OECD system for monitoring and controlling transfrontier movements of hazardous wastes is compatible with existing international agreements governing transport of dangerous goods. The prior notification information referred to under (c) includes data concerning which (if any) of the transport of dangerous goods protocols the wastes will be carried. Moreover, packaging of the wastes for transport and classification concerning transport hazards must conform with United Nations recommendations (United Nations, 1988). The proposed OECD hazardous waste notification and shipment documentation in no way supplants or interferes with documentation demanded by any existing international agreement concerning transport of dangerous goods (OECD, 1989).

Wastes subject to the proposed OECD control system must be classified by means of an International Waste Identification Code (IWIC) which was formally agreed by member countries in May 1988 (OECD, 1988). The IWIC allows virtually all wastes deemed to be hazardous by most countries to be described satisfactorily in terms of potential hazard, activity generating the wastes, physical form (liquid, sludge, solid), generic descriptor (contaminated soil, solvent mixture, etc.) and constituents. In addition, the IWIC indicates reasons why the wastes were intended for disposal plus the disposal operation(s) to which the wastes will be subjected—whether recycling or 'final' disposition. The IWIC provides for a 'cradle-go-grave' dossier for any batch of wastes. The IWIC is coded against a set of master tables and is thus essentially independent of language.

The OECD draft International Agreement does not deal directly with liability issues which may arise in case of mishap during transfrontier movement of hazardous wastes. The reasons for this situation are that a draft Convention on Civil Liability for Damage Caused During Carriage of Dangerous Goods by Road, Rail and Inland Navigation Vessels (the so-called CRTD Convention) was adopted on 10 October 1989 (United Nations, 1989*a*). A close counterpart would deal with liability for damage caused during carriage of dangerous goods on the high seas, i.e. the so-called HNS Convention.

The CRTD proposals are intended to cover all transfrontier movements of hazardous wastes and, among other things, apply to:

(a) loss or damage to the environment caused by the dangerous goods (or wastes), provided that compensation for impairment of the environment other than for loss of profit from such impairment shall be limited to costs of reasonable measures of reinstatement actually undertaken or to be undertaken;
(b) the costs of preventive measures and further loss or damage caused by preventive measures wherever such measures are taken; and
(c) all operations from the beginning of loading the goods (or wastes) onto the vehicle for carriage until the end of operations of unloading the goods (or wastes).

Who is liable and how to ensure that liability claims are properly taken into account provide the main substance of the CRTD proposals. In brief, liability is usually channelled toward the carrier. Compulsory insurance is required of carriers so that funds are available for mitigating any claims found valid under the terms of the CRTD. The HNS is likely to operate in an analogous manner.

Proposals are thus in a very advanced stage to provide a clear internationally binding scheme to address liability issues for transfrontier movements of hazardous wastes from loading through all carriage operations to completion of unloading. Domestic legislation, regulations or policies governing hazardous waste liability in the country of importation would apply fully once the wastes were unloaded.

The CRTD and HNS Conventions, once implemented, will provide the basis for dealing with liability issues, which may arise during transfrontier movement of hazardous wastes. This being the case, no pressing need to address liability issues directly in the OECD draft International Agreement existed. Indeed, any such efforts would have duplicated the CRTD and HNS work at best and could have confused matters at worst.

The OECD draft International Agreement takes into account the possibility that hazardous wastes may be destined for disposal in an area outside national jurisdiction, for example, a marine area. In that case, the country of importation is taken to be the country where transfer to the vessel which is designated to bring these wastes to the disposal area occurs. The authorities of that country would be provided with the notification referred to in (c) above (OECD, 1989).

European Community

The European Community (EC) has adopted a series of Directives aimed at monitoring and controlling transfrontier movements of hazardous wastes. Intracommunity movements as well as exports from or imports into the EC have been taken into account by these Directives. In developing these Directives, the Commission of the European Communities (CEC), having collaborated closely in similar OECD efforts, elected to adapt certain results achieved in the OECD to EC needs.

A Directive on the supervision and control within the European Community of the transfrontier shipment of hazardous waste was adopted in December 1984 (84/631/EEC). This Directive largely transmuted the 'Principles' included with OECD Council Decision–Recommendation C(83)180(Final) into a legally binding form for use within the EC. This Directive excludes controls for chlorinated and organic solvents and provides for minimal controls with respect to waste, scrap, sludge, ash and dust from non-ferrous metals which is intended for re-use, regeneration or recycling on the basis of a contractual agreement regarding such operations (European Community, 1984).

Directive 84/631/EEC was supplemented in October 1985 by a further Directive providing details of precisely how the notification of transfrontier movements of hazardous wastes must be done in the EC. In particular, specific forms and procedures for notification were laid down (European Community, 1985).

The EC method of control of transfrontier movements of hazardous wastes was to have come into use in the autumn of 1985. However, delays were encountered for a variety of reasons. (Perhaps a major cause was that no two EC countries had identical means of specifying what constituted a hazardous waste nor were lists of such wastes the same in any two countries.) These difficulties were summarized by the relevant EC Commissioner in late 1988 (Pienaar, 1988). At present, the EC control method is largely operating among EC countries.

Directive 86/279/EEC sets forth conditions governing control of exports of hazardous wastes from the EC area (European Community, 1986). This Directive closely resembles its earlier OECD counterpart described earlier.

Recently, the CEC has proposed two new Directives which would define 'wastes', 'disposal' and 'hazardous wastes' within the EC (European Community, 1988*a,b*). These proposals were developed after close collaboration of the CEC in the OECD negotiations which led to adoption by the OECD Council of the Decision of 27 May 1988 described on page 144.

United Nations Environment Program

In June 1987 the Governing Council of the United Nations Environment Program (UNEP) decided to proceed with development of a global Convention concerning the control of transboundary movements of hazardous wastes. Similar work in OECD was specifically mentioned by the Governing Council; results achieved at OECD were to provide a foundation insofar as practicable for the UNEP efforts (United Nations, 1987*a*). The UNEP Secretariat adopted a very ambitious schedule; a Convention was adapted on 22 March 1989 (United Nations, 1989*b*).

The goals set by UNEP go beyond effective monitoring and controlling transfrontier movements of hazardous wastes. Environmentally sound management of hazardous wastes is also taken into account. Not surprisingly, very close attention is being paid to identifying clearly the problems and needs of developing countries in these areas (see note 3). A key objective of the Convention is to help these problems and to prevent developing countries from becoming repositories for improperly identified and improperly managed hazardous wastes. Another key objective is to pinpoint clearly what constitutes illegal traffic (United Nations, 1987*b*) in hazardous wastes and to create a means to punish perpetrators of such traffic. Yet another key objective is to ensure mechanisms for redress in case of illegal or inappropriate exports of hazardous wastes to developing countries.

There is considerable difference of opinion and debate concerning precisely what approaches should be adopted for minimizing the possibility of inappropriate transfer of potentially hazardous wastes from industrialized to developing countries. There are advocates for a total ban of such exports, i.e. all transfers of any wastes proscribed as 'hazardous' toward developing countries would be illegal and thus punishable under terms of the UNEP Convention. Some proponents of such a ban would be prepared to allow transfers if bilateral or multilateral agreements between governments setting out specific terms were negotiated and adopted. Others suggest that a system for orderly control of transfrontier movements of hazardous wastes is preferable to an outright ban. Not all of the 'ban' advocates are from developing countries, and not all of the 'orderly control' advocates are from industrialized countries.

One of the compromises considered in the UNEP negotiations included the possibility of a codicil to the Convention whereby any country which had prohibited the import of all 'wastes' (see note 4) or 'hazardous wastes' would be listed. Prospective exporting countries would agree not to allow movements of hazardous wastes to be initiated towards countries included in this list. The EC and 68 African, Caribbean and Pacific (ACP) countries have since agreed that EC Member states shall ban all exports of hazardous and nuclear wastes to the ACP countries. In turn, the ACP countries shall not import such wastes from any other non-EC country.

Another possibility is to require that prior notification of hazardous wastes which are proposed for entry into the territory of a developing country—either for transit or for disposal—be provided on a standard Certificate of Origin (CoO) form. The CoO is commonly used in international export and import operations and so should be familiar to customs officers and other officials concerned with trade. Often, a Consular official of the country into whose territory a shipment is to enter must sign the CoO before the shipment commences. Thus, this approach, if applied to wastes, would provide for prior informed consent—in written form as attested by a Consular official—*prior to legal commencement* of a proposed transfrontier movement of hazardous wastes. Clearly, Consulates would consult with their appropriate national authorities before electing to sign or to refuse to sign the CoO. This approach provides a means to reduce the probability of stranded shipments of hazardous wastes, e.g. the well-publicized *Karin B, Zenoobia* and other similar cases.

Existing OECD and EC requirements are such that a transfrontier movement of hazardous wastes destined for a developing country should be stopped by the exporting country unless an appropriate official of the importing country has provided written consent to the import *and* unless the wastes are directed toward an 'adequate disposal facility'. For the UNEP Convention, the term 'adequate disposal facility' could perhaps be defined as 'something built, installed or established to perform hazardous waste disposal operations in a manner such that any and all requirements pertaining to these operations as published in legislation, regulations, policies or licensing rules in the country of exportation are met'.

This definition does not infringe upon the sovereignty of any country of importation nor does it require the country of exportation to enforce any laws but its own. Rather, this definition sets the same standards of care for wastes to be exported as for wastes to be managed 'at home'. Exporting countries would require that whoever must notify the country of importation inform the relevant authorities of the standards which must be met in order for the export to proceed without objection. In addition, exporters would be required to provide information confirming, in fact, that the facility proposed for disposal in the country of importation can meet these standards.

Adoption of a stringent definition of 'an adequate disposal facility' might cause concerns that legitimate recycling/reclamation activities in developing countries would be deprived of feedstocks. On the other hand, sham recycling schemes (see note 5) might be curtailed or eliminated. Indeed, several of the spring 1988 press reports concerning waste shipments to developing countries intimated that sham recycling was involved.

For wastes destined for recycling/reclamation operations, 'an adequate disposal facility' might be one that would be eligible (authorized or allowed) or licensed to perform these operations if it were located in the country of exportation. The notifier would inform the relevant authorities of the country of importation of the standards which must be met in order for the export to proceed without objection. In addition, exporters would be required to provide to the competent authorities information confirming that the facility proposing to perform the recycling/reclamation operations can, in fact, meet these standards.

While these steps may seem a bit draconian, their implementation would merely confirm that potentially hazardous wastes need a high degree of monitoring and control. Many OECD member countries have defined such controls by means of domestic laws and regulations. There is empirical evidence that developing countries do not want to import wastes if they will be managed less stringently than required under the laws of the country of exportation. Countries of exportation are in the best position to know what standards of waste management need to be applied to wastes generated in their country. If these standards cannot be met, the inescapable conclusion is that the potential for harm to man and/or the environment from the wastes is in some way increased.

Those who negotiated the text of the Basel Convention [United

Nations, 1989*b*] grappled with these and a number of other issues—such as liability, insurance, illegal traffic—in an effort to develop a strong basis to monitor and control transfrontier movements of hazardous wastes. Though the Convention was adopted on 22 March 1989, when full and effective implementation of the UNEP global system will occur is difficult to predict at present.

PROGNOSTICATIONS CONCERNING THE NEAR-TERM FUTURE

At present, certain potentially hazardous wastes—if they are not legally defined as or considered to be hazardous wastes by the exporting country—need not be notified in advance if (and only if) they are destined for disposal operations leading to resource recovery, recycling, reclamation, direct re-use or alternative uses. If such potentially hazardous wastes are directed toward an importing country which has few or even no control mechanisms for dealing with these wastes, then the possibility for sham recycling is quite strong. A number of developing countries are well aware of this fact.

Certain countries, such as the Dominican Republic, have simply elected to legally ban the import of wastes. Other countries may not wish to ban imports but rather to receive notice of *all* wastes which are proposed for entry. Thus, certain countries may well choose to enact domestic legislation requiring that all imports of items referred to as 'wastes' as delineated by the UNEP Convention be notified in advance.

Some countries are proposing the definitions and listings included in OECD Council Decision C(88)90(Final) (OECD, 1988) as the basis for their respective domestic legislation concerning management of wastes. Presumably, these countries intend to develop domestic monitoring, control and management schemes for all such wastes. The draft UNEP Convention contains provisions which would enable such countries to request and to receive advice and assistance in implementing these schemes.

For hazardous wastes proposed for import into developing countries, suggestions have been made that private sector firms could be employed to verify empirically that notification information concerning wastes to be imported is identical with the actual wastes which have arrived in the customs-bond area in the country of importation. If not, the shipment would be rejected and would then perhaps be considered as a transfrontier movement which could not be completed as foreseen. In this case, the wastes would be eligible for repatriation at the expense of the exporter. (N.B. Presumably, the exporter would be expected to pre-pay the estimated costs of the empirical verification and perhaps would also be expected to post a bond covering costs of repatriation if the shipment is, in fact, rejected.)

The OECD Environment Committee considered in April 1989 the implications and modalities of further restricting hazardous waste exports to developing countries pending entry into force of a global Convention. A wide variety of options and their probable impacts if adopted were taken into account.

Ratification and full implementation of the Basel Convention will not occur until sometime in the future. In the interim, OECD-Member country governments have pledged to 'make special efforts to ensure that existing legal and administrative instruments for controlling transfrontier movements of hazardous wastes are applied, consistent with the intent of the draft OECD International Agreement and the draft global Convention. In so doing, particular attention will be given to measures intended to protect public health and the environment in developing countries' (OECD, 1989*a,b*).

In 1989, the bases for effective monitoring and control of transfrontier movements of hazardous wastes were put into place. Existing instruments such as OECD legislation, EC Directives, International Agreements governing transport of dangerous goods, and domestic laws, regulations and policies appear to be sufficient for monitoring and control of these movements among industrialized countries such as OECD and Council of Mutual Economic Assistance members. The global Convention, when fully implemented and enforced, is meant to ensure that effective and appropriate monitoring and control of transfrontier movements of hazardous wastes occurs worldwide. One caveat is necessary, however. There must be active, strong and effective domestic waste management systems in force. If not, an international system may be built on a foundation of sand. The domestic control system *must* mesh with the international system in order to provide maximum protection for man and/or the environment.

NOTES

1. The Organization for Economic Cooperation and Development (OECD) is concerned with helping to achieve high economic growth, employment and a rising standard of living while at the same time ensuring that environmental amenities are maintained or improved and that man and the environment are protected insofar as possible from harm. In other words, the OECD actively seeks to promote sustainable growth. Policies and practices concerning hazardous waste management differ in each OECD member country as compared to the others.

   Member countries of OECD include Australia, Austria, Belgium, Canada, Denmark, Finland, France, Germany, Greece, Iceland, Ireland, Italy, Japan, Luxembourg, the Netherlands, New Zealand, Norway, Portugal, Spain, Sweden, Switzerland, Turkey, United Kingdom, United States. Yugoslavia participates in certain work including environmental issues. Note that all 12 countries of the European Community are also OECD members.

   In 1974, the OECD Environment Committee, which guides all work involving environmental matters on behalf of the member countries, created the Waste Management Policy Group. This Group is composed of high level governmental officials charged with responsibility for achieving environmentally sound waste management in their respective countries. As its name implies, the Group is primarily concerned with considering, developing and promulgating international policy to promote appropriate waste management as a contributor to sustainable growth. In its work, the Group considers technical, economic and social issues and seeks to harmonize results of its work into coherent and implementable international policy actions.
2. The Council subsequently extended this time limit to 31 December 1988.
3. The Annex contains brief background comments concerning hazardous waste management issues faced by developing countries.
4. As defined by the UNEP Convention. At present, the proposals for these definitions are virtually identical to those adopted by OECD Member countries.
5. If hazardous wastes are directed toward inappropriate, inadequate or even non-existent resource recovery, recycling, re-

clamation, direct re-use or alternative use operations or activities, then the term "sham recycling" is often employed to describe this situation.

REFERENCES

European Community (1984). Council Directive on the supervision and control within the European Community of the transfrontier shipment of hazardous waste. 84/631/EEC, 8 December.

European Community (1985). Council Directive adapting to technical progress Council Directive 84/63/EEC. 85/469/EEC, 22 July.

European Community (1986). Council Directive on the export of hazardous waste from the European Community. 86/279/EEC, 12 June.

European Community (1988*a*). Proposal for a Council Directive amending Directive 75/442/EEC on waste. 88/C295/03. *Official Journal of the European Communities,* C295/3–C295/7, 19 November.

European Community (1988*b*). Proposal for a Council Directive on hazardous waste. 88/C295/04. *Official Journal of the European Communities,* C295/8–C295/16, 19 November.

Levine, A. G. (1982). *Love Canal: Science, Politics and People.* Lexington Books, Lexington, MA, 263 pp.

OECD (1984). Transfrontier movements of hazardous waste. Decision and Recommendation of the Council, C(83)180(Final), adopted 1 February.

OECD (1985). International cooperation concerning transfrontier movements of hazardous waste. Resolution of the Council, C(85)100, adopted 20 June.

OECD (1986). Exports of hazardous wastes from the OECD area. Decision/Recommendation of the Council, C(86)64(Final), adopted 5 June.

OECD (1988). Transfrontier movements of hazardous wastes. Decision of the Council, C(88)90(Final), adopted 27 May.

OECD (1989*a*). 'Control of Transfrontier Movements of Hazardous Wastes'. (Resolution of the Council), (C(89)1), adopted 27 January, 1989.

OECD (1989*b*). 'Control of Transfrontier Movements of Hazardous Wastes'. (Resolution of the Council), [C(89)112(Final)] adopted 20 July 1989.

Philips, A. & Silbergeld, E. (1984). 'Comments of the Environmental Defense Fund on 'Health Aspects of the Disposal of Waste Chemicals' Prepared for Chemical Manufacturers Association (U.S.) by Universities Associated for Research and Education in Pathology'. Environmental Defense Fund, Washington, D.C., 17 December, 1984.

Pienaar, J. (1988) 'EC Commissioner Attacks Delays in Curbing Toxic Waste', Independent (U.K.) for 29 November, 1988.

Reko, H. K. (1984). *Not an act of God: The Story of Times Beach.* Copyright 1984 by H. K. Reko, 80 pp. plus Appendix.

United Nations (1987*a*) [Environment Programme]. Report of the Governing Council-Decision 14/30 on 'Environmentally Sound Management of Hazardous Wastes" dated 17 June, 1987.

United Nations (1987*b*) [General Assembly]. Resolution 42/183 defining 'Illegal Traffic'.
United Nations (1988). [Economic and Social Council]. Recommendations on the Transport of Dangerous Goods. (Fifth revised edition), ST/SG/AC. 10/1/Rev. 5, UN–New York, 1988, 504 pp.
United Nations (1989*a*). [Economic Commission for Europe—Inland Transport Committee]. 'Convention on Civil Liability for Damage Caused During Carriage of Dangerous Goods by Road, Rail and Inland Navigation Vessels (CRTD)', Geneva, 10 October 1989.
United Nations (1989*b*). [United Nations Environment Programme]. 'Basel Convention on the Control of Transboundary Movements of Hazardous Wastes and Their Disposal', Basel, 22 March 1989.
Yakowitz, H. (1984). 'Economic and Policy Issues Associated with Transfrontier Movement of Hazardous Waste'. Unpublished report to the OECD Waste Management Policy Group (dated 17 February, 1984), 14 pp.
Young, B. A. (1983). Statement before the Subcommittee on Commerce, Transportation and Tourism of the Committee on Energy and Commerce, U.S. House of Representatives: Implementation of the Superfund Program, 22 November, 1983, serial No. 98–128, pp. 266–268.

APPENDIX: SOME COMMENTS CONCERNING HAZARDOUS WASTE MANAGEMENT ISSUES FACED BY DEVELOPING COUNTRIES

(Excerpted from a report prepared in 1985 by the author at the request of the World Commission on Environment and Development.)

Broadly speaking, three primary sources of hazardous wastes in developing countries exist:

(a) wastes generated within the country either by foreign-owned, state-owned or joint-venture firms;
(b) wastes imported into the country; and
(c) wastes generated within the country by small entrepreneurs such as electroplaters, metal finishers, etc., as well as local farmers and householders.

Waste amounts in all of these categories are likely to increase rapidly in the next 10–20 years as reference to OECD registers of development research projects and technology transfer data clearly indicate. The hardest category to control for the developing countries is likely to be that listed as (c) above.

Until recently there were few figures detailing the export of hazardous wastes from developed to developing countries. (But then, perhaps, such data were not likely to be reported.) Such traffic can be very profitable and may well increase in the future. Many developing countries are experiencing rapid industrialization and a corresponding increase in hazardous wastes. Very few, however, have any legislation concerning the management of hazardous wastes. There is usually no policy regime, no institutional capacity and essentially no resources to devote to the issue. The potential for waste tourism, both north to south and south to south thus exists and is likely to grow. Developing countries must become aware of the potential dangers of such traffic in order to deal properly with transfrontier movements of hazardous wastes

Some countries already have, and other countries may be tempted to, become 'buyers' and even 'traders' in hazardous waste. Given comparative levels of development, awareness and information flows, they may be able to sustain such trade for some time. In the medium to longer term, however, such trade is not likely to be satisfactory either economically or politically.

Each improperly managed disposal site may reduce the future development potential of a country. In the longer term, when development in the importing countries leads to higher levels of education, information flows and awareness in the communities suffering the consequences of such trade, the exporting countries concerned run the risk of being charged with taking advantage of unequal relations (neo-colonialism) and perhaps worse.

Economically, such trade makes no sense for the buyer, except in terms of the most primitive short-term financial analysis. Improper disposal can end up costing a society 100–1000 times more than environmentally sound management of the wastes. Neither does trade make any long-term sense for the seller, but rather lowers the pressure to find environmentally and economically acceptable ways of managing the waste within the industries, communities, countries and regions generating it.

Note that the types of wastes most likely to be candidates for export are highly hazardous, i.e. wastes which are very expensive to eliminate in the home country because of their nature (highly toxic, corrosive, etc.) and which are banned from legal dumping into the sea. Hence, importing countries should be able to manage such wastes properly or they may run a very serious risk of adverse effects on man and/or the environment.

According to UNEP findings, waste management in developing countries suffers from a variety of difficulties. For example, countries in or near the tropical rain belt must cope with frequent heavy rains; landfilled waste is thus subject to the strong possibility of rapid leaching of even direct overflow due to water content. Since there may be little or no pretreatment of the waste prior to land-filling, this practice could lead to contamination of water supplies (ground and surface water) as well as to direct exposure of nearby dwellers. Again, according to UNEP, most industrial production in developing countries is concentrated in congested areas. Land-filling of hazardous waste generally occurs close to industrial estates which are surrounded by poor neighbourhoods. These dangers point up the clear need for land use planning in developing countries and the more urgent need to implement and enforce such plans.

Hazardous wastes can also lead to severe problems for animals and marine life in groundwater and surface water contaminated by wastes. This problem is especially worrisome in areas with porous soil and subject to torrential rains. Run-off from hazardous waste sites can also contaminate large areas of land. In communities strongly dependent on agriculture such an occurrence, which could result in large amounts of land being rendered useless, might be catastrophic.

In developing economies, the state may be unable to successfully appropriate enough resources from internal sources to fuel industrial and commercial growth without strongly affecting ingrained social and cultural customs. Thus, in many instances, the state apparatus with respect to control over the activities of possible sources of hazardous wastes is neither well-developed nor strong.

This situation means that there is a possibility for certain enterprises to seek short-term benefits to themselves by lowest cost disposition of hazardous wastes. Such disposition may not always be environmentally appropriate. Thus, the governments of many developing economies may well be faced with a future need to react and cure uncontrolled disposal areas. Technical expertise and considerable funds may be required for such tasks. The alternative to remedial action could be serious harm to man and/or the environment.

Therefore, the key issue for developing economies is how to institute some means of monitoring and control of hazardous wastes *now* without major new bureaucracies or expenditures of funds being required. In other words, what 'anticipate and prevent' strategy can be applied most effectively at minimum cost given the existing situation in developing economies? Note that this strategy should also include export–import considerations, i.e. transfrontier movements of hazardous wastes must be taken into account. Finally, when uncontrolled sites from past practices are suspected to be present, some means of planning for eventually dealing with such sites, e.g. developing funds to apply to the problem, deserves consideration.

The components for a system to monitor and control hazardous wastes in developing economies may ultimately take a variety of forms depending upon country, waste arisings and local institutional arrangements. Nevertheless, certain features which might be common to a number of monitoring and control systems in developing economies can be suggested.

(1) Information exchange between governmental entities and the generators of hazardous wastes is necessary. Government must inform generators of plans and requirements with respect to management of hazardous wastes. The first step is to promulgate a list of waste substances which are to be included in the control system. This list should be developed in consultation with the generators; the list should be as simple as possible. Lists of enterprises from which potentially hazardous wastes may arise are available to provide guidance to authorities as to whom potential generators of such wastes may be.

(2) Codes of practice for hazardous waste management could be developed and agreed between generators and government authorities. Developing economies could probably implement codes of practice with the cooperation of the regulated community. Existing site licences for co-disposal of urban waste and special wastes could be examined for cues as to how to proceed in order to achieve low-cost environmentally sound disposition of hazardous wastes.

Many developing economies are moving toward environmental assessment as one of the tenets for accepting expanded or new industrial development. Some form of environmental audit suggests itself for the control of hazardous wastes since such wastes have the potential to seriously harm man and/or the environment. This audit procedure can be part and parcel of the environmental impact assessment. The policy issues include who shall perform the audit, how frequently will the audit be repeated, who will have access to the results since industrial secrets may be involved, and who shall pay for the audit? (One could conceive of the generators pooling resources to pay qualified auditors subject to a secrecy agreement. The government would select an auditor from the approved list, pay from the pool and only the government and the audited party would have access to the data. Many other approaches are possible, of course.)

(3) Proper hazardous waste management must be ensured for expanding or new development of enterprises which generate such wastes. Contractual obligations are usually agreed between investors and host country governments. Requirements that investors, i.e. generators of hazardous wastes, dispose of their wastes as in the home country could be imposed. Indeed, certain large multinational chemical firms already assume such a burden. Thus, achieving agreement along these lines might be possible. The assumption is that auditing will be included as part of the contractual obligations and that hazardous wastes are handled appropriately in the home country.

Moreover, when special treatment of waste is required, the investor should agree to provide for such treatment. Independent experts could be utilized to advise both parties as to the type and cost of such special arrangements so as to protect man and/or the environment at lowest practicable cost.

(4) Funds must be readily available to deal with possible mishaps associated with hazardous wastes. Few developing economies have the means to deal with a major environmental problem requiring costly remedial action. Hazardous waste mishaps usually occur infrequently, but the cost of appropriate action to deal with such mishaps properly can be enormous. Who may have responsibility to provide for remedial action and how is a key policy point.

(5) Other issues. (i) Education and training. Monitoring and control of hazardous wastes is relatively straightforward to discuss in theoretical terms. But in the real world, a certain level of education, training and awareness is needed with respect to the personnel who must oversee the operation of the system. Developed economies often do not have the cadre of trained and motivated personnel required. The situation in developing economies is somewhat worse.

While guidelines and regulations can be imposed by governments, they must be put into practice correctly and equitably by appropriate personnel. Developing economies often do not have the specialized educational training curricula for such waste management officials. Nor may the resources and opportunity exist to train personnel abroad.

Hence, a crucial requirement for developing economies is some means to obtain local personnel who are fully trained in waste management. An appropriate form of assistance either from MNE generators of hazardous wastes or their host governments might be the development, implementation and maintenance of educational efforts meant to provide host developing economies with the necessary cadre of trained personnel. In the absence of such personnel, no waste management system is likely to be very effective.

(ii) Many developing economies have not prepared an overall waste management scheme. This issue is a corollary to the education and training problem. Development and implementation of an overall waste management scheme dealing with all types of solid, sludgy and liquid wastes (including wastewater) is not likely without trained and highly motivated local personnel who have the full backing of their government. Calling in an expert consultant may be necessary but is unlikely to be sufficient for developing and implementing an effective plan geared to local conditions and customs.

(iii) Regional harmonization of waste management might be effective in many areas. If one country in a given area takes effective steps locally with respect to hazardous waste management, its efforts may be discounted by transboundary pollution from neighbouring countries. For example, groundwater contaminated by hazardous wastes in country 'X' might be used ultimately for drinking purposes in country 'Y'.

Thus, regional councils, cooperation and mutual help are very likely to be necessary in many of the developing economy areas of the world. Rapid organization and deployment of such groups could be very useful in minimizing the chance of mishap (and costs associated therewith) due to improper hazardous waste management.

In sum, the development of guidelines or codes of good practice or whatever one wishes to call well-intentioned and necessary documents is a useful first step for realizing appropriate waste management. But there can be no substitute for the presence of trained and motivated personnel who can adjust to local conditions and customs in order to ensure that appropriate waste management occurs in practice.

Therefore, policy goals must almost certainly include people-oriented waste managers as well as guidelines and schemes to obtain resources. The combination of these two units may provide sufficient ingredients to realize means for waste management in developing economies which will protect man and/or the environment.

An important requirement in many developing countries is to recruit, train and put in place senior policy personnel who can determine the issues the country faces with respect to hazardous waste management, and advise on and implement policies in the context of the myriad of other problems and priorities faced by the country.

Technical expertise, e.g. waste site operators, is of secondary importance at present.

This task requires resources and funding, something which few developing countries are ready to assign to this priority. One possibility, therefore, in the cases of wastes of MNE origin, is to tie funding for training and engaging management personnel to the investment that gives rise to the wastes that need to be managed. Furthermore, where the financing of an enterprise is assisted by bilateral or multilateral agencies, they may be asked to reinforce such a policy and facilitate training. Industry-sponsored non-governmental organizations may also be effective in providing such training.

Given developing countries' priorities, resources for hazardous waste management, and environmental management generally are and will remain extremely scarce. Strengthened international cooperation at the regional level could greatly increase the effectiveness of all resources provided from whatever source. An appropriately mandated and structured regional body, for example, could provide policy advisory services to the countries in the region; it could suggest legislation, policy and regulations; it could receive 'notification' of shipments on behalf of countries not equipped to handle them and advise on whether or under what conditions a country should 'consent' to a shipment of waste; it could advise on negotiations with new industries that will produce new wastes. Such bodies could produce contingency plans and manage contingency resources. It could maintain a regional quick-response capacity for use in case of accidents due to improper hazardous waste management. Such a body, moreover, could be the overseer and administrator of funds earmarked for hazardous waste management in the region. The pooling of talent and resources in this way could also increase awareness and strengthen the political capacity of governments in a region to deal with both external and internal sources of hazardous waste.

13

Reducing Corporate Environmental Liabilities with a Complete Environmental Management and Protection System: A Guide for Directors, Officers and Managers of Manufacturing Corporations on How to Avoid Environmental Suits and Charges

RONALD A. N. MCLEAN

Arthur D. Little Ltd, Science Park, Cambridge CB4 4DH, UK

and

MARIE-CLAUDE SAVOIE

Arthur D. Little Inc. (Canada), 67 Yonge Street, Toronto, Ontario, Canada

In recent years, environmental legislation around the world has put the onus on individuals, as well as corporations, to prevent pollution damage from manufacturing operations. The United States has led the way in putting the liability on individuals in its environmental legislation and US courts have liberally chosen to hold corporate officers, managers, etc. liable, whether or not the corporation is found liable. In Scandinavia, corporate managers can be found criminally liable for transgressions of environmental statutes for which the corporation had some responsibility and of which they, as individuals, had some knowledge. Other countries have followed suit by ensuring that individuals, as well as corporations, can be charged with environmental offences of which they had some knowledge and in making environmental infractions criminal offences. The US Superfund program under the Comprehensive Environmental Response, Compensation and Liability Act (CERCLA) has made various parties strictly and retroactively liable for the cost of remedial actions at selected waste sites even if sites were approved, transporters licensed and wastes registered. However, this has now been taken a step further and legislation is being passed where directors, officers and managers of corporations can be found *criminally* and strictly liable for a pollution offence, even though they were not aware of the pollution act or its background, the corollary being that they should have been aware.

A particular example of this is the Environmental Enforcement Statute Law Amendment Act 1986, which was brought into force on 18 December 1986 in the province of Ontario, Canada. Provisions were added to that province's Environmental Protection Act defining the personal liability of directors and officers and, incidentally, drastically increasing fines and adding jail sentences for offences. The relevant provisions in Section 147A of the Environmental Protection Act are as follows:

(1) Every director or officer of a corporation that engages in activity that may result in the deposit, addition, emission or discharge of a contaminant into the natural environment contrary to this act or its regulations has a duty to take all reasonable care to prevent the corporation from causing or permitting such unlawful deposit, addition, emission or discharge.
(2) Every person who has a duty under Subsection 1 and who fails to carry out that duty is guilty of an offence.
(3) A director or officer of a corporation is liable to conviction under this section whether or not the corporation has been prosecuted or convicted.

The legal implications of this section and other recent amendments to the Environmental Protection Act of Ontario have been discussed in detail at several environmental law conferences over the past 2 years and corporations have been reviewing in considerable detail their own environmental management systems. Particularly useful, for guidance in this review, is the paper by Cotton & Nicholls (1988), on which the extended recommendations in this paper are based. As might be expected, some directors view the onus on them under this legislation as so heavy that they have resigned or considered resigning. Some legal firms have requested that their partners, who could be affected by this legislation should not sit on the boards of manufacturing companies, for fear of incurring unexpected liabilities.

Corporations are increasingly indemnifying or providing assurance for their officers and directors against the possibility of prosecution under environmental legislation. It is strongly recommended that corporations obtain separate coverage for pollution in which they would list officers' and directors' names, but even this extra coverage would not cover officers and directors under shareholder suits. It is uncertain what view the courts will take of this kind of coverage and in fact this may encourage higher penalties. Of course there is no insurance policy against jail sentences! As such, directors are now required to take a much more active role in ensuring that their company has an adequate environmental management system and complies with environmental law.

The spirit of Sections 147A of the Ontario Environmental Protection Act clearly requires that directors and officers (in this case, officers could be any responsible person above a first line supervisor), *themselves,*

(a) maintain themselves current with the environmental risks in the corporation;
(b) ensure that experts advise them on the risks and means to reduce them;
(c) monitor regularly efforts to evaluate and reduce environmental liabilities; and
(d) review carefully regular environmental reports from the corporation.

It is now an important duty of all directors and officers to continually satisfy themselves that the corporation is meeting (or is aiming to meet, with adequate regularly reviewed action plans) all its environmental obligations, e.g. is not significantly polluting, has eliminated all risks of environmental accidents and has safe products. If acts of pollution etc. are brought to their attention, directors, officers or managers must take all reasonable steps to prevent them continuing or recurring. In an atmosphere of 'guilty until proven innocent' the directors and officers must arm themselves with all the due-diligence equipment necessary to ensure that they reduce environmental risks. An officer or director who does not believe that he is capable of getting the information necessary to protect himself, should seriously look at his position: quoting John D. Honsberger (1988): 'The new rule for directors and officers concerning environmental hazards is *caveat diribetor*—let the director beware'.

The question is often asked what type of corporation is at high risk of investigation by agency enforcement officers or of being charged with environmental offences. The following five criteria are worth consideration:

(1) *A negative public environmental profile of the corporation.* This is a corporation which frequently gets negative national or local press on environmental issues, e.g. frequent spills, association with downstream drinking water or fish contamination or with beach closings, complaints from its neighbours of gaseous emissions, odours or particulate fall-out, particularly with negative health implications, contaminated sites to clean up.
(2) *A senior corporation.* A well-established, profitable company is more likely to be a target than a small, less profitable one because the enforcement agent's attitude is likely to be that the well-established company should have known better.
(3) *Poor environmental track record.* The corporation which has a poor reputation with the agency in terms of lack of cooperation with the agency's abatement or enforcement officers, fails to meet agreed abatement deadlines, has consistent confrontations with the agency and treats the agency's officers as the enemy, is likely to be a better target than if it has a record of good cooperation with the agency.
(4) *A level of responsibility for directors, officers and managers.* Those who are the most senior in the decision-making line are most likely to be charged.
(5) *A poor environmental management and protection system.* A corporation with a poorly documented incomplete environmental management and protection system is more likely to be a target than one which has a very well-documented system.

This paper attempts to arm the officers and directors in their efforts to establish in their companies as complete a system of environmental management and protection as possible. This is a guide for the reduction of the risk of liabilities of environmental offences in on-going operating plants for directors, officers and operators, risk and environmental managers in manufacturing industries, large and small, in any jurisdiction throughout the world. This paper does not deal with liabilities related to past practices and resulting on-site and off-site contamination and remediation. We necessarily take a pro-active position. No longer is it satisfactory for corporations, independent of their size or apparent risks, to take the position that they shall wait until the environmental agency tells them to do something before doing it. It is essential for adequate protection that *all* manufacturing (and many service) corporations, not just those with apparently the highest risks such as the largest international commodity chemical companies, put in place a complete environmental management and protection system of the type outlined in this paper. We believe that this type of system is essential for manufacturing corporations whose operations do or may discharge to the environment and we strongly recommend that they be pro-active and ahead of regulatory requirements. It is also important for corporations who do not have discharges to institute the relevant parts of the system, since there are many environmental laws and regulations which apply to them, even though there may be no obvious pollution damage they could cause. The items, which together form our Complete Environmental Management and Protection System, can be summarized under ten sections as follows:

(1) Environmental expertise
(2) Environmental policy and procedures
(3) Environmental law and regulatory advisory program
(4) Environmental audit and compliance action program
(5) Environmentally friendly processes and products program
(6) Environmental protection program
(7) Emission reduction program
(8) Emergency prevention contingency planning, and community awareness
(9) Duties of individuals and training requirements
(10) Reporting to officers and the board of directors

A brief summary will be given of each item.

As an example to illustrate the environmental requirements in each section, we assume that we are dealing with Corporation A which uses gas X to weatherproof the natural organic material M to produce MX (a fibre for clothing) in a number of plants around the world. Gas X is regulated by the agencies because it is acutely toxic by inhalation to humans and because it reacts with natural organic matter to produce a reaction by-product XY known to cause an unusual cancer in dogs. The plants must use water to wash MX and the solids from the wastewater must be settled. The wastewater in some plants is partly or fully recycled into the process and, at other plants, is discharged for treatment to an industrial treatment system or directly to receiving waters. The waste solids are sent to hazardous waste landfills off-site.

No other substances used or produced, other than X and XY, are regulated, although there are literature references to damages to fish caused by other by-products of the treatment of M by X. There are no known significant environmental or human health hazards related to MX.

To begin each section we ask a key question which directors must pose to officers and officers to their environmental experts to assure themselves that adequate programs are in place or will be put in place.

(1) Environmental Expertise

Question
May I meet the key individual(s) with responsibility for the corporation's environmental programs?

Summary
Whether a manufacturing company is large or small, has a centralized or highly decentralized management style, it must have access to the appropriate environmental expertise. Every corporation should have reporting to the Chief Executive Officer or its Chief Operating Officer and, preferably to the head of each large operating unit, a full-time environmental manager and/or environmental legal counsel (for large corporations) or an external environmental consultant with a wide general knowledge of the field (for small, low-risk corporations) who would set up and implement the Complete Environmental Management and Protection System. The environmental manager or consultant must have appropriate resources to put in place the policy and procedures of Section 2 and all the programs identified in Sections 3–8 below. The environmental lawyer(s) should be consulted on a regular basis on new laws, regulations and guidelines (henceforth regulations) and, as needed, for applicable existing laws and regulations.

Example
Corporation A should have among others an environmental manager or consultant who is knowledgeable or has access to experts in (a) the inhalation toxicology of gas X, (b) assessment of risks of accidental emissions of X and its effects on the surrounding population and methods to reduce the risks, (c) environmental chemistry of the by-products of the reaction between M and X and the effects of these on municipal treatment systems and on biota in receiving waters, (d)

hydrogeology to understand the movement of XY and other by-products from landfills, and (e) the toxicology to humans of MX, XY and its by-products through drinking water, and physical contact, in addition to the ecotoxicology for biota in the receiving waters.

(2) Environmental Policy and Procedures

Question
Does the corporation have a policy on environmental matters?

Summary
Every corporation should have a clear written policy to direct its operations and its activities in environmental matters. This policy should cover, in general terms, the requirements for each of the sections (3–8) following in this list and from it should stem detailed, clearly written procedures to cover all requirements for good environmental management.

Example
Corporation A would have extensive procedures covering the handling of X, and the measurement of discharges and their impact, and control of MX, XY and others. Such procedures should be equally strong at all plants.

(3) Environmental Law and Regulatory Advisory Program

Question
How do our operations remain up-to-date on applicable environmental regulations?

Summary
All operations of the corporation should provide themselves or be provided by the corporate department with regular updates on environmental regulations and, in particular, be warned *in advance* of any upcoming regulations. The operations should be fully conversant with the requirements of environmental permits, licences, orders, etc. The programs should be in place to meet and maintain compliance with these requirements. Operating staff should be trained in the appropriate environmental regulations affecting their operations. There should be regular consultation on environmental regulations

with an environmental counsel, knowledgeable in the jurisdictions of the operations.

Example
The plant manager or his delegate should be familiar with all regulations on X and XY in that jurisdiction. The corporate manager and/or his staff should be familiar with *all* regulations on X and XY in the jurisdictions where corporation A has plants *and* in all other jurisdictions where tighter standards are applied.

(4) Environmental Audit and Compliance Action Program

Question
May I have a review of the corporation's environmental audit program?

Summary
There should be in place a comprehensive environmental audit program carried out independently from operations to assess and verify the compliance status of operations with present and known future regulations, with company policy and procedures, and with good environmental management practices and to ensure that an efficient environmental management system is functioning properly. Of even more importance, the environmental audit program should have associated with it a follow-up compliance action program to carry out the corrective actions and other recommendations of the audit. It is essential that the board of directors and senior management institute these programs, that they monitor them on a regular basis and that the findings and action plans be reported to and approved by them. This will constitute a necessary part of a due-diligence defence if anything does go wrong.

If environmental auditing is a 'voluntary necessity' as headlined in an editorial in *Water and Pollution Control* in July 1988 (Anon., 1988), the compliance action program to correct audit findings is absolutely essential. Auditing and compliance action programs are basically good business strategy. Not only do they identify potentially costly environment risks and speed up compliance, they also help to improve the overall management system of an operation. They can also help to significantly improve productivity and efficiency and reduce costs.

Example
Corporation A should have a comprehensive audit program. Because of the nature of the raw material, X and the by-products, all plants should be measured against the toughest regulatory standards at a minimum.

(5) Environmentally Friendly Processes and Products

Question
How do we plan to reduce exposure to hazardous chemicals and to convert to environmentally friendly products and processes?

Summary
Every corporation should have programs in place (a) to evaluate the risks to its workers, the environment and the public of its raw materials, products, by-products, (particularly potentially hazardous chemicals) and processes, (b) to reduce and eliminate the use, exposure to and discharge of the hazardous chemicals, and (c) to reduce the hazards both in processes and in products to the environment and individuals. All corporations should have a long-term strategic plan to convert to environmentally friendly processes and products.

Example
Corporation A should have a research and development program in place internally or with a consultant with the aim of replacing gas X as a treatment agent for the weatherproofing of material M with a less hazardous material, which would not be acutely toxic to humans and which would not produce by-products which are carcinogenic to animals.

(6) Environmental Protection Program

Question
What program(s) does the corporation have in place to ensure environmental protection?

Summary
Every corporation should establish and operate effectively an environ-

mental protection program, which assures that the company's operations have minimal environmental impact and comply with all present appropriate environmental legislation and will comply with future legislation. The Environmental Protection Program should include at least the following six elements.

(a) monitoring and inventorying amounts of raw materials, by-products, products, and discharges and concentrations and amounts of contaminants in the discharges from operations to the atmosphere, groundwater and surface water, soils, in wastes leaving the site and in products;
(b) Assessment of the impact of these materials and discharges on the environment and consumers;
(c) compliance of operations with appropriate environmental regulations and corporate policy, particularly with respect to the discharge of contaminants, control programs and measures to prevent continuous pollution and accidental spills,
(d) maintenance of all pollution control and other process equipment with a regular preventative maintenance program;
(e) action plans to meet and maintain compliance with regulations and corporate policy and procedures and to reduce the impact of operations on the receiving environments; and
(f) control programs to prevent any unwanted discharges.

It is essential that a corporation has a detailed knowledge of all of its discharges and their contaminants and has programs in place to reduce such discharges to meet the strictest regulatory requirements and the standards set by the corporation's own impact assessment studies. The corporation's long-term strategic planning should include the planning of the environmental protection program. This planning should not only include the reduction and elimination of discharges from operations, but also should plan to reduce the use of or eliminate all hazardous chemicals used in the process.

Example
Corporation A should have an extensive program of monitoring discharges of gas X and by-product XY (most likely required by regulations) and any other by-products of the reaction between M and X which it feels could have a significant impact on the environment. Certainly, it should have sufficient monitoring data on the dis-

charges of X to the atmosphere and of XY to receiving waters and in solid or liquid wastes, to establish with confidence the average daily discharge and the range of daily discharges of these materials. From the literature information on the effects of X on humans, it should be able to establish the effect of its discharges on their own workers and the surrounding population. It should be investigating the levels of the other by-products of the treatment of M by X and determining whether the levels of these by-products in receiving waters are anywhere near the levels determined to have caused damage to fish. It should be carrying out or have carried out studies of the effects of the discharges of these materials on receiving waters.

Compliance with regulations for discharges of X and XY are essential but, in addition, the corporation should be setting its own standards for the discharges of X and XY from all of its plants to meet the toughest regulations, provided these are possible on a technical basis. The regular preventative maintenance program should focus on or should have a high focus on the handling of the dangerous gas X. Action plans should be drawn up to reduce discharges of X and XY to receiving environments especially if there is difficulty meeting regulations or the company's own standards. Finally, the company's longer term planning should include investigations to replace X by some other material in the weatherproofing of MX (see Section 5).

(7) Emission Reduction Program

Question

Based on existing regulations and the company's assessment program (Section 5b above), do we have a long-term plan to control emissions of contaminants to receiving environments?

Summary

The corporation should have in place a long-term plan and detailed programs to control the emissions of contaminants to receiving environments, water, air, soil, in products and in wastes being moved from the plant site. This program should cover both continuous discharges and any potential accidental discharges. The initial basis for the program will be existing and forthcoming proposed regulations. However, the company's own environmental protection program may indicate requirements for control, over and above existing regulations.

Example
Corporation A should have programs in place to further control emissions of X to the atmosphere and of XY and other by-products of the treatment of MX, to water and wastes. Wastewater control and recycling programs should be in place for all operations and some methods should be found not to produce solid wastes containing the contaminant XY.

(8) Emergency Prevention, Contingency Planning and Community Awareness

Question
How do we maintain and update our emergency prevention and contingency plans? Are the emergency prevention plans based on adequate risk assessments for all possible environmental accidents?

Summary
Every corporation should have all of its operations prepare adequate emergency prevention and contigency plans (a) to reduce the risk of significant accidents taking place, and (b) to ensure the proper response and reporting is provided if an emergency does occur.

If the emergency prevention plan is not based on complete and adequate risk assessments of potential accidents for all key operations in the plant or if adequate equipment is not in place to prevent spills, or the serious effects of spills, this must be planned in a timely manner. Continuous training should be given to all staff in the plant on the use and operation of such environmental emergencies and contingency plans. The plans should be reviewed for completeness by experts from outside of the plant.

Example
Corporation A's emergency prevention and contingency planning will focus highly on preventing any emergencies involving gas X. Every aspect of X's use from the time it leaves the supplier until it is used up in the plant should be examined in detail to determine the risk of some accident. This would include detailed risk assessment of transportation to the plant, unloading, storage, use in the process, recovery of the material, and environmental control systems, using Hazop or some other risk assessment technique. This would be followed by detailed

recommendations on preventative measures to ensure that no such release can take place. In addition to the risk reduction studies, a detailed examination of all possible accident scenarios should be carried out in order that complete satisfactory evacuation plans are prepared.

(9) Duties of Individuals and Training Requirements

Question
How are the duties of each individual in the corporation involved in environmental matters defined and what training does he/she receive?

Summary
For each individual in the corporation involved in environmental matters up to and including officers and directors, the duties with respect to this Environmental Management and Protection System should be clearly spelled out in writing. Training programs with regular refresher courses should be put in place for all participants. This training should be ongoing with regular refresher courses. Performance on meeting present defined goals, in meeting objectives of the programs in the system should be evaluated for the annual salary increase, bonus, etc.

(10) Reports to Officers and the Board of Directors

Question
How are relevant changes in environmental regulations, the results of environmental audits, any situations of significant non-compliance in the plants and any potential charges or lawsuits reported to the Senior Management and the Board of Directors?

Summary
Regular (monthly or quarterly) reports should be made to the officers and sent from them to the board on relevant changes in environmental regulations, results of environmental audits, any situations of non-compliance in the plants and any potential charges or lawsuits. Officers and the board must review all environmental reports carefully, act diligently to take the necessary actions to correct any significant areas of non-compliance reported and to deal, in a timely manner, with any reports of environmental concern from the public, employees, govern-

ment agencies, etc. If initial indications of environmental concern are dealt with immediately and the appropriate corrections are made, this will help the company avoid enforcement action from the agencies in the future.

Most corporations, with known or potential environmental concerns will already have in place significant portions of the above Environmental Management and Protection System. Much of it will be part of their normal operating practice. We believe that all that is suggested above is now part of good management practice in the modern industrial world. Although initially there may be some significant costs with introducing some of these items, most companies who have taken a pro-active position on environmental matters have discovered significant unexpected pay-backs in the long run. Therefore, the above programs are recommended, not only to protect the environment and reduce the liabilities for the corporation, its officers, directors and managers, but to improve operating efficiency.

REFERENCES

Anon. (1988). Editorial. *Water and Pollution Control,* July.

Cotton, R. & Nicholls, Lori J. (1988). Minimizing pollution offence liability of Corporate directors and officers. Paper presented at the Conference entitled The Company Polluted—So Why Did I Get Charged?, 21 January. Canadian Environmental Law Research Foundation.

Honsberger, D. (1988). Paper presented at the Conference entitled The Company Polluted—So Why Did I Get Charged?, 21 January. Canadian Environmental Law Research Foundation.

14

Immobilization of Pollutants in Stabilized Sludge Composed of Lignite Ashes and Residues of Flue Gas Desulphurization

Hans Ulrich Bambauer, Christian Krause, Gudrun Willner

Institut für Mineralogie, Universität, Correnssstraße 24, D-4400 Münster, FRG

Georg Gebhard, Thomas Holzapfel

Rheinisch-Westfälisches Elektrizitätswerk, Kruppstraße 5, D-4300 Essen, FRG

and

Walter Lukas

Institut für Baustofflehre und Materialprüfung, Universität, Technikerstraße 13, A-6020 Innsbruck, Austria

EXTENDED ABSTRACT

The lignite-fired power plants of the Rhineland produce about 6·1 million t of lignite ashes, 1·3 million t of gypsum from the flue gas desulphurization (FGD), and 750 000 m^3 of 'FGD-water' (circulation water enriched in trace elements and chlorides ranging up to 10 g/litre Cl^-), per year.

In order to minimize groundwater contamination the following concept (besides technical barriers in the dump) has been developed: mixing of appropriate amounts of ashes and gypsum (as far as not recycled) with diluted 'FGD-water' (water/solid ratio approximately 0·3) to form a sludge ready for disposal. As a result of mineral reactions the paste-like mixture sets in the disposal dump to give a solid material similar to a supersulphated slag cement mortar. Analogous laboratory samples (unit weight 1·7–2·0 g/cm^3) show relatively high compressive strength (5–15 N/mm^2; 180 days), low water permeability ($k = 2 \times 10^{-9}$–1×10^{-10} m/s) and leachability.

The raw ash is mainly composed of quartz, iron oxides, coke particles, glassy microspheres, CaO, MgO, Ca-ferrite, (aluminates), calcium sulphate and various accessory minerals. From these starting materials neomineralization starts at an initial temperature of about 40°C—accompanied by increasing condensation of microtexture—to form portlandite, brucite, gypsum, calcium silicate hydrates, phases of the ettringite and 'monosulphoaluminate' type and thaumasite as well as calcite and various neoformed accessories. Quartz, coke, iron oxides and glassy microspheres form the fine aggregate.

Among the very minute ($\theta < 1$–$10\ \mu m$) neoformed minerals, in general, portlandite and ettringite (rarely thaumasite) could be detected by X-ray powder methods, whereas the remaining minerals were mostly determined by analytical electron microscopy.

Condensation of microtexture (and reaction rate) is distinctly increased by raising the chloride content of the water. As a function of duration the chloride content is rapidly used up: easily soluble chlorides decrease whereas slightly soluble Cl-containing sulphate–silicate ettringite phases (in part also 'monosulphoaluminate' phases) increase rapidly. In addition, the ettringite phases incorporate besides Cl^- (and NO_3^-) into their crystal structures considerable amounts of heavy elements like Cr, Mn, Fe, Co, Ni, Zn, Pb and As. Cadmium was found in 'monosulphoaluminate' only. These form a first generation of slightly soluble 'reservoir minerals' which effectively immobilize the pollutants mentioned above. Also, heavy metal compounds form (mostly) as slightly soluble accessories; for instance Sr-containing barite, Zn-containing brucite, Zn hydroxide and Mn hydroxide. The more slowly forming calcium silicate hydrates, as a secondary generation of 'reservoir minerals', also cause a further lowering of porosity.

Hence, condensation of microtexture (lowering of porosity) and crystallochemical immobilization of pollutants as a result of neomineralization is apparently related to decreasing permeability and considerable lowering of chloride-, nitrate-, and heavy metal leachability. Samples prepared in the laboratory are not essentially different from those taken as drilling cores from the disposal dump.

An overview of the presented waste management concept is given by Gebhard *et al.* (1989), and some fundamentals of the mineralogical part appeared recently (Bambauer *et al.*, 1988*a,b*).

REFERENCES

Bambauer, H. U., Gebhard, G., Holzapfel, Th., Krause, Chr. & Willner, G. (1988*a*). Schadstoff-Immobilisierung in Stabilisaten aus Braunkohlenaschen und REA-Produkten. I. Mineralreaktionen und Gefügeentwicklung; Chlorid-Fixierung. *Fortschr. Miner.*, **66**(2), 253–79.

Bambauer, H. U., Gebhard, G., Holzapfel, Th. & Krause, Chr. (1988*b*). Schadstoff-Immobilisierung in Stabilisaten aus Braunkohlenaschen und REA-Produkten. II. Schwermetall-Fixierung, Bilanzen. *Fortschr. Miner.*, **66**(2), 281–90.

Gebhard, G., Bambauer, H. U. & Lukas, W. (1989). Stabilisierung von Rückstanden aus der Braunkohleentschwefelung. *VGB-Kraftwerkstechnik*, **69**(4).

Appendix I

International Society for Environmental Protection (ISEP)

FOUNDATION AND OBJECTIVES

The International Society for Environmental Protection (ISEP) was founded in 1987, as a result of an initiative by an international group of experts. ISEP is a non-profit association of a new kind, a forum for the presentation of different views on environmental policies at an international level, with special emphasis on fair and equal cooperation between science and industry.

ISEP constitutes an independent platform for all countries and their organizations, for science and industry, technology and management, research and production, concerned with environmental protection on an international scale. The International Society for Environmental Protection is based in Vienna, a city well-known as a venue for international meetings and as a centre for worldwide cooperation.

To further ISEP's major interest in promoting environmental management activities on an international scale, a special sponsoring membership is offered to industrial companies and organizations.

CURRENT ACTIVITIES

On 20–22 February 1989, the First International ISEP Congress, Envirotech Vienna 1989, offered to more than 1000 participants from 42 countries the opportunity of high-level discussions on scientific, technical and economic aspects of urgent environmental protection problems.

Envirotech Vienna 1990 is planned for 23–25 October 1990 at the Austria Center, Vienna, with special emphasis on 'Current Problems in Hazardous Waste Management and Contaminated Sites'. The conference will take place together with the international scientific fair, World Tech Vienna 1990. From 1990 onwards, Envirotech Vienna will take place each year at the end of October.

Another ISEP event is the 'International Environment Days Bad Kleinkirchheim' (Carinthia), which is held as an annual seminar. In 1989 invited participants and ISEP members gathered on 24–26 September to discuss 'Technology, Innovation and European Market Economy'.

ISEP's international activities are supported by a network of National Points of Contact (NPOCs), maintaining close cooperation and promoting the exchange of information with the Vienna-based ISEP secretariat. ISEP events may be organized by NPOCs in collaboration with the international secretariat. Organizations and individual experts interested in cooperating with ISEP as National Points of Contact are invited to address their proposals to the ISEP board.

For more information, please contact the ISEP secretariat:

International Society for Environmental Protection
A-1090 Vienna
Spittelauer Lände 3
Austria

Telephone: 43/222/34 45 11 510
Telefax: 43/222/34 45 11 299

Appendix II

ENVIROTECH VIENNA 1989 – Kongreßberichte / Proceedings

SESSION 1: *BEURTEILUNG UMWELTGEFÄHRDENDER STOFFE UND PRODUKTE*

SESSION 1: *EVALUATION OF HAZARDOUS SUBSTANCES AND PRODUCTS*

(erhältlich bei/available at: Westarp Wissenschaften, Segerothstr. 73, D – 4300 Essen 1)

1 TOXIKOLOGISCHE ASPEKTE DER BEWERTUNG UMWELTGEFÄHRDENDER STOFFE UND PRODUKTE

1 ASSESSMENT OF HAZARDOUS SUBSTANCES AND PRODUCTS: TOXICOLOGICAL ASPECTS

Risk Assessment of Pulmonary Carcinomata Caused by Radon Exposure
K. KARRER, Univ. of Vienna, A

Zur epidemiologischen Bewertung von Karzinogenen
M. KUNDI, M. NEUBERGER, Univ. Wien, A

Zu einigen chemisch–toxikologischen Aspekten der Chemikalienbewertung
K. LOHS, Akad. d. Wiss., Leipzig, DDR

Schätzung des nitratbedingten Risikos in einem Struma–Endemiegebiet – Ergebnisse einer epidemiologischen Studie
H. HÖRING, M. NAGEL, Forschungsinst. f. Hygiene u. Mikrobiologie, Bad Elster, DDR

Evaluation of Carcinogenic Potential of Synthetic Rubbers Containing Mineral Oils
M. RATPAN, Polysar Ltd, Sarnia/Ont., A. HAYES, Univ. of Guelph, CDN

2 ÜBERWACHUNG UND BEWERTUNG UMWELTGEFÄHRDENDER CHEMIKALIEN

2 CONTROL AND ASSESSMENT OF ENVIRONMENTAL CHEMICALS

Strategien und weitere Entwicklung des deutschen Chemikalienwesens im Hinblick auf eine umfassende Umweltpolitik
U. SCHLOTTMANN, Bundesministerium f. Umwelt, Naturschutz u. Reaktorsicherheit, Bonn, D

A Theoretical and Computational Approach to Chemical Evaluation Based on Structure–Activity Relationships
G.J. NIEMI, S. BASAK, Univ. of Minnesota, G.D. VEITH, EPA, USA

Calculations of Releases and Environmental Concentrations in the Risk Assessment for New Chemicals
P.v.d. POEL, J. ROS, National Inst. for Public Health and Environmental Protection, Bilthoven, NL

Sorption of Fire Extinguishing Agent CF_3Br by PVC
S.H. HÜTTENHAIN, R. FELDMANN, Fraunhofer–Inst. f. Umweltchemie u. Ökotoxikologie, Schmallenberg, D

Derzeitige Grenzen und Entwicklungsbedarf von Bewertungsverfahren für umweltgefährdende Stoffe und Produkte
H.M. KNOFLACHER, G. RETTENSTEINER, Forschungsges. Joanneum, Graz, A

Access and Review of Data for Environmental Chemicals
K. VOIGT, J. BENZ, R. BRÜGGEMANN, M. MATTHIES, GSF München, Neuherberg, D

3 SICHERHEITSKONZEPTE IM PRODUKTIONS– UND FORSCHUNGSBEREICH

3 SAFETY CONCEPTS IN INDUSTRIAL PRODUCTION AND RESEARCH

Structural Measures as a Preventive Answer to Environmental Catastrophes
H. BOMHARD, Dyckerhoff & Widmann AG, Munich, D

Safety in Biotechnological Research with Special Emphasis on Plant Premises
B. MÜLLER, ETH Zürich, CH

4 ENTWICKLUNGSTENDENZEN UND PROBLEME DER SCHADSTOFFANALYTIK

4 NEW DEVELOPMENTS AND CURRENT PROBLEMS IN ENVIRONMENTAL ANALYSIS

European Projects on Sulphur Dioxide Reference Materials for Environmental Analysis
E. GOELEN, T. RYMEN, Nuclear Research Center, Mol, B

Analysis of Aromatic Hydrocarbons in Overburden of Coal Mining: Assessment of Environmental Risks
W. PÜTTMANN, RWTH Aachen, D

Umweltkontrolle auf chlorierte Dioxine und Furane in der Umgebung einer Metallrückgewinnungsanlage in Tirol
A. RISS, Umweltbundesamt, Wien, A, H. HAGENMAIER, Univ. Tübingen, D, W. ROTARD, Bundesgesundheitsamt, Berlin, D

TECAN CRANOX–System – Korrekte Analyse von kleinen NO– und NO_2–Immissionskonzentrationen
D.G. RYHINER, Tecan AG, Hombrechtikon, CH

Kraftwerk Dürnrohr – Ein neues Umgebungsüberwachungssystem
H. LINHART, W. DAXBÖCK, Verbundkraft GmbH, Wien, F. HIRSCHMANN, EVN AG, Maria Enzersdorf, A

High–Sensitivity Measurements of Wet and Dry Deposition in Venice
A. RIGO, Univ. of Padova, N. BASILI, G.R. GUIDOTTI, ENEA, Roma, I

New Developments in Determination of Very Low Concentrations of Nitrogen Oxides (NO_x) in the Ambient Air by Spectrophotometric Methods
P. SAGNER, Chemoprojekt, Praha, CS

Radon in Dwellings: Investigations and Measurements
F. STEGER, H. STADTMANN, P. KINDL, Austrian Research Centre Seibersdorf / TU Graz, A

Determination of Low Concentrations of Chlorinated Hydrocarbons Round Vinylchloride–Monomer Plants by Gas–Phase Infrared Spectrophotometric Multicomponent Method
L. KACELE, Chemoprojekt, Praha, CS

Analytical Evaluation (GC and NMR Methods) of Microbial Ecosystem Potential in Off-Shore Environment Polluted by Oil-Refinery Products
M. VRDOLJAK, I. PAVUSEK, F. ANUSIC, G. PROTIC-LOVASIC, J. MÜHL, INA / Univ. of Zagreb, YU

Development of a General Analytical Scheme for Organic Chemicals in Soil
U. WAHLE, W. KÖRDEL, W. KLEIN, Fraunhofer-Inst. f. Umweltchemie u. Ökotoxikologie, Schmallenberg, D

Grundlagenerfassung für die Schadstoffausbreitung – Bisherige Ergebnisse aus der Steiermark
R. LAZAR, Univ. Graz, A

Determination of Some Minor and Trace Elements in Fly Ash, Wet-Filter Water and Wet-Filter Cake of an Incineration Plant by Short-Time Activation Analysis
F. GRASS, G.P. WESTPHAL, S.S. ISMAIL, F. BENSCH, Atominst. d. Österr. Universitäten, H. LÖFFLER, MA 22, Vienna, A

Total Reflection X-Ray Fluorescence Analysis – A Method for Environmental Analysis
P. WOBRAUSCHEK, H. AIGINGER, C. STRELI, P. KREGSAMER, Atominst. d. Österr. Universitäten, Vienna, A

5 BEWERTUNG UMWELTGEFÄHRDENDER STOFFE UND PRODUKTE – ANGRENZENDE GEBIETE
5 ASSESSMENT OF HAZARDOUS SUBSTANCES AND PRODUCTS – RELATED SUBJECTS

Monitoring for Potential Environmental Problems at a Nuclear Research Facility at Hanford, Washington, U.S.A.
R.H. GRAY, Pacific Northwest Laboratory, Richland/WA, USA

Erfassung von Latschenbeständen (Pinus Mugo Sp.) im Karwendel-Gebirge mit Hilfe von CIR-Luftbildern
J. HACKL, E. KNAPPITSCH, Umweltbundesamt, Wien, A

Ökotoxikologische Bewertung des SANDOZ-Schadenfalles
U. MATTHIAS, Landesanstalt f. Umweltschutz Baden-Württemberg, Karlsruhe, D

Limit Values for Establishing Hazard and Hazard Classification of Wastes and their Applicability
A. HORVATH, H. FARKAS, National Inst. of Public Health, Budapest, H

Eco Packages, New Ecological Programmes for Food Packing
J. GVOZDENOVICH, M. CURAKOVIC, V. LAZIC, I. VUJKOVIC, Faculty of Technology, Novi Sad, YU

Critical Levels of Pollution from the Viewpoint of Corrosion
D. KNOTKOVA, K. BARTON, P. HOLLER, G.V. Akimov Research Inst., Praha, CS

The ÖNORM S 5200 "Radioactivity in Building Materials" as a Tool for Radiation Protection of the General Population
B. KUNSCH, F. STEGER, Austrian Research Centre Seibersdorf, E. TSCHIRF, Atomic Inst. of the Austrian Univ., Vienna, A

Die Umweltbelastung in der Türkei durch die Verbrennung uranhaltiger Braunkohlen und Asphaltite als Folge ihrer Lagerstättenbedingungen
H. NEWESELY, M. ANDAC, TU Berlin, D

Transfer of Soluble Constituents from a Subsiding Ash Deposit to Water at the "Sostanj" Power Plant
B. STROPNIK, K. JUZNIC, P. STEGNAR, Jozef Stefan Inst., Ljubljana, YU

SESSION 2: *AKTUELLE ENTWICKLUNGEN IN DER UMWELTSCHUTZTECHNIK*

SESSION 2: *CURRENT TRENDS IN ENVIRONMENTAL PROTECTION TECHNOLOGIES*

(erhältlich bei/available at: Westarp Wissenschaften, Segerothstr. 73, D – 4300 Essen 1)

1 ABLUFT– UND ABGASREINIGUNGSVERFAHREN

1 WASTE AIR AND WASTE GAS DECONTAMINATION TECHNOLOGIES

The 3 S – Process: Simultaneous (Acid Gas and NO_x) Sodium–Based Sorption of Flue Gases
R. LINDBAUER, M. BOBIK, Waagner–Biro AG, Graz, A

Removal of Trace Metals at High Temperatures
W. MOJTAHEDI, Technical Research Centre of Finland, Espoo, SF

Environmental Pollution by Odorous Emissions of Domestic and Toxic Waste Landfills – Basics, Present Situation and Reduction by Waste Gas Treatment Techniques
F.–B. FRECHEN, Hydro–Ingenieure GmbH, Düsseldorf, D

Fortschrittliche Rauchgasreinigung nach Verbrennungsanlagen für Klärschlamm und Sonderabfall
H. HOFSTETTER, EBS, Wien, A

Ionizing Wet Scrubbers – Advanced Air Pollution Control Systems for Hazardous Waste Incineration
M. GRIEM, Ceilcote GmbH, Biebesheim, D

Continuously Cleaned Cross–Flow Electrostatic Precipitator
H. JODEIT, E. FLECK, Asea Brown Boveri, Baden, CH

Deodorization and Purification of Waste Air by Ozone
C. DARPIN, Asea Brown Boveri, Zurich, CH

Characterization of the Flow Reactor for the Determination of the SO_2 and HCl Absorption in Limestone and Dolomite
M. LODDER, Univ. Hannover, D

Entwicklung eines biologischen Abluftreinigungsverfahrens für einen Viskosebetrieb
A. WINDSPERGER, A. SCHMIDT, TU Wien, L. KLOIMSTEIN, Glanzstoff AG, St. Pölten, E. NIEDERMAYER, L. BERZACZY, Waagner–Biro AG, Wien, A

Die Abluftreinigung bei der Holzspantrocknung
A. SCHMIDT, TU Wien, A

Removal of Sulfur Dioxide from Waste Gases by a Sodium Citrate Solution
B. SKRBIC, J. CVEJANOV, Fac. of Technology, Novi Sad, YU

2 VERFAHREN DER ABWASSERREINIGUNG

2 SEWAGE TREATMENT TECHNOLOGIES

Cobalt Recovery from Waste Water by Liquid Surfactant Membranes
I. ABOU–NEMEH, A.P. van PETHEGHEM, Univ. of Gent, B

Überwachung von Kläranlagenzuläufen zum Schutz vor toxischen Abwasserinhaltsstoffen
M. EWALD, Fachhochschule Niederrhein, Krefeld, D

Components of an Advanced Biological Sewage Purification
M. BOBIK, A. GLASNER, G. BECKMANN, Waagner–Biro AG, Graz, A

Solvent Extraction – An Efficient Tool for Waste Water Treatment
H.J. BART, R. MARR, TU Graz, A

Thermische Oxidation von Salz– und Organikahaltigen Abwässern
H.P. JAEGER, Schering AG, Wolfenbüttel, D

Behandlung feindisperser Systeme durch Elektrokoagulation
E. GIDARAKOS, V. FRIEHMELT, A. KÖHLING, R.E. SCHMITT, Battelle Inst., Frankfurt, D

Aufbereitung von Abwässern aus der Rückstandsentsorgung mittels Querstrommembranfiltration
J. GNIESER, A. GMÜNDER, Sulzer AG, Winterthur, CH

Die Sauerstoffversorgung der Biozönose in Belebungsanlagen. Der Stand der Technik von Belüftungseinrichtungen
H.J. SCHMIDT–HOLTHAUSEN, Suprafilt GmbH, Bietigheim, D

Method of the Water–Oil–Emulsions Elimination in the Environmental Protection
L. TOKARZEWSKI, J. BOREK, B. KOZBIAL, Silesian Univ., Katowice, PL

Biotechnologie bei der Behandlung von Abwässern aus industrieller Produktion und Abfalldeponien
P. BALDIA, H. WAGINGER, Wien, A

3 TECHNOLOGIEN DER ABFALLENTSORGUNG
3 WASTE DISPOSAL TECHNOLOGIES

The Immobilization of Pollutants in Stabilized Sludge composed of Lignite Ashes and Residues of Flue Gas Desulphurization
H.U. BAMBAUER, C. KRAUSE, G. WILLNER, Univ. Münster, D,G. GEBHARD, T. HOLZAPFEL, RWE, Essen, D, W. LUKAS, TU Innsbruck, A

Weitergehende Rauchgasreinigung durch Kondensation–Naßelektrofilter. – Erste Betriebserfahrungen hinter einer Sonderabfallverbrennungsanlage
K.v. BECKERATH, Bayr. Landesamt f. Umweltschutz, München, D

A Thermal Process for the Detoxification of Filter Ash from Waste Incinerators
M. HIRTH, J. JOCHUM, H. JODEIT, C. WIECKERT, Asea Brown Boveri, Baden, CH

Recent Advances in Solidification/Stabilization Technologies
R. SOUNDARARAJAN, RMC Environmental Laboratories, West Plains / Mont., USA

The Effect of Oxygen Enhancement on Hazardous Waste Incineration
L. STALEY, US Environmental Protection Agency, Cincinnati / Ohio, USA

Verbrennung PCB–haltiger Altöle im Zementdrehofen
G. KROGBEUMKER, Phoenix Zementwerke, Beckum, D

Jarosit als Thallium–Träger in einer Deponie von Schwefelkiesabbränden
W. STEFFES–TUN, H.U. BAMBAUER, Univ. Münster, C. KRAUSE, Kernforschungszentrum, Karlsruhe, D

The Treatment of Hazardous Wastes from Lithium Battery Production
J. ZERJAV, Jozef Stefan Inst., Ljubljana, YU

4 UMWELTGERECHTE RECYCLING–VERFAHREN
4 LOW POLLUTING RECYCLING TECHNOLOGIES

Anfall und Entsorgung von Reststoffen aus der Rauchgasreinigung
K.H. GRUBER, R. HAMMERSCHMID, R. WINKLER, O. RENTZ, TH Karlsruhe, D

The DUT Low Temperature Regenerative Desulfurization Process in Fluidized Bed Combustion of Coal
A.E. DUISTERWINKEL, E.H.P. WOLFF, E.B.M. DOESBURG, C.M. van den BLEEK, TU Delft, NL

Möglichkeiten der Entsorgung metallhaltiger Abfallstoffe und Altlasten
W. JANKE, D. HANKEL, Lurgi GmbH, H. MACZEK, Metallges. AG, Frankfurt, D

The Long Heralded Breakthrough for Re–Refining of Used Lubricating Oil – Theory and Reality
F. SAGER, Fellow Inst. of Petroleum (London), Consultant, S

The Tubular Reactor: A New Process for Reclaiming Used Oil and Decontaminating Other Hazardous Wastes
C. SCHÖN, ENTRA GmbH, Achern, P. WALDT, SÜDÖL GmbH, Eislingen, D

Separation of Zinc, Cadmium and Lead by a Combined Extraction/ Liquid Membrane Process
J. DRAXLER, H.J. BART, R. MARR, TU Graz, A

Economical Recovery of High Quality Polyethylene from Mixed Plastic Waste
G. HÖRBER, AKW GmbH, Hirschau, D

5 AKTUELLE KONZEPTE "SAUBERER TECHNOLOGIEN"
5 NEW TRENDS IN "CLEAN TECHNOLOGIES"

Clean Technologies Policy of the European Economic Community
E. MURILLO MATILLA, Commission of the European Communities, Brussels, B

State and Trends of Low–Waste Technologies in the Italian Industry
G.R. GUIDOTTI, A.M. CASTAGNOLA, ENEA, Rome, I

Wastewater–Free Metal Plating
K. SKOV, National Agency of Environmental Protection, Copenhagen, DK

A Model Photoelectrochemical Process with n–CdS Particles Immobilized in Gel Membranes
A.M. MIKA, TU Wroclaw, PL

Technological and Economical Prospects for Low Emission Energy Systems
V. OKOROKOV, IIASA, Laxenburg, A

Potential Uses of Fly Ash as Valuable Material in FGD and Some Other Applications
A. STERGARSEK, Edvard Kardelj Univ., Ljubljana, YU

SESSION 3: *SONDERABFALL UND ALTLASTEN*

SESSION 3: *HAZARDOUS WASTE AND CONTAMINATED SITES*

(erhältlich bei/available at: Westarp Wissenschaften, Segerothstr. 73, D – 4300 Essen 1)

1 NEUE KONZEPTE FÜR ABFALLWIRTSCHAFT UND GEFÄHRLICHE SONDERABFÄLLE

1 NEW POLICIES IN WASTE AND HAZARDOUS WASTE MANAGEMENT

Stand der Sondermüllentsorgung in Bayern unter besonderer Berücksichtigung thermischer Entsorgungsverfahren
W. RUCKDESCHEL, Bayr. Landesamt f. Umweltschutz, München, D

Strategie zur Lagerung von Abfällen nach den Leitlinien zur Abfallwirtschaft in Österreich
P. LECHNER, TU Wien, A

Abfallwirtschaft und Altlastensanierung – Ein wichtiger Schwerpunkt der österreichischen Umweltschutzgesetzgebung
W. LIST, Bundesministerium für Umwelt, Jugend und Familie, Wien, A

Transfrontier Movements of Hazardous Wastes: The UNEP– and OECD–Draft Conventions – A Comparative Study of Legal Approaches to Manage an International Problem
C. STIX–HACKL, Federal Ministry of Foreign Affairs, Vienna, A

An Approach of Prevention and Recycling of Waste
M.M.J. ALESSIE, Ministry of Housing, Physical Planning and Environment, Leidschendam, NL

Risk Assessment in Hazardous Waste Remediation
J.D. LEE, A.J. VALDES, SRI International, Menlo Park/CA, USA

Entwicklung eines Prognose– und Entscheidungsmodells der Abfallentsorgung
W. PILLMANN, A. HOLZLEITNER, ÖBIG, Wien, A

Abfallentsorgung – Eine Bürde und eine Chance für die Unternehmen
E.v. PERFALL, Ruhrkohle Umwelt GmbH, Essen, D

Das Branchenkonzept – Ein neuer Ansatz zur Lösung abfallwirtschaftlicher Probleme aus Kleinbetrieben
B. BILITEWSKI, intecus, Berlin, D

Gefährdungspotentiale der relevanten Abfallarten und Bedeutung der Untertagedeponie als Entsorgungseinrichtung
G. JAKOB, Inst. f. Tieflagerung, GSF, Braunschweig, D

2 BEWERTUNG DES GEFÄHRDUNGSPOTENTIALS VON ALTLASTEN

2 RISK ASSESSMENT OF CONTAMINATED SITES

Erkennung, Charakterisierung und Bewertung von Grundwasserbeeinflussungen durch Abfallablagerungen
R. SCHLEYER, G. MILDE, H. KERNDORFF, J.–D. ARNETH, Bundesgesundheitsamt, Berlin, D

Methodological and Policy–Related Problems and Opportunities in the Assessment of Risks from Contaminated Sites
T. ASSMUTH, M. MELANEN, National Board of Waters and the Environment, Helsinki, SF

Erkundung kontaminierter Standorte, sogenannter Altlasten mit geophysikalischen Methoden
R. BACHER, ÖMV AG, Gänserndorf, A

Expertensystem zur Bewertung des Gefahrenpotentials von Altlasten
H.–P. LÜHR, Inst. f. wassergefährdende Stoffe a.d. TU Berlin, D

ALTRISK – Ein wissensbasiertes System zur Bewertung des Gefährdungspfades Grundwasser
G. OSTERKAMP, B. RICHTER, W. SKALA, FU Berlin, D

Development of Risk Assessment and Management Method for Planning Landfill Disposal Site
T. FURUICHI, M. TANAKA, Inst. of Public Health, Tokyo, J

Die Untertagedeponie – Ein Konzept zur Ablagerung chemisch–toxischer Abfälle
T. BRASSER, T. MEYER, C. STARKE, Inst. f. Tieflagerung, GSF, Braunschweig, D

Simulation of Groundwater Flow in Crystalline Formations with the Use of a Three–Dimensional, Dual–Porosity Model
H. KULL, GSF, Braunschweig, J. LEITENBAUER, Texaco AG, Hamburg, J. SIEBERT, R. WEBER, TU Clausthal, D

Schwermetallgehalte und Auslaugungsverhalten verschiedener Bodenarten im Hamburger Raum
A. BAERMANN, Geologisches Landesamt Hamburg, D

Altlasten–Erkundung mit Hilfe einer bildgebenden, elektro–magnetischen Sondierung
V. GENRICH, Genitron Instruments, G. SCHMIDT, Intellex Expertensysteme, Frankfurt, D

Tiedebeeinflußte Verlagerung von Schadstoffen im überkonsolidierten, geklüfteten Ton
A. BAERMANN, Geologisches Landesamt Hamburg, J. BRUNS, INHAK Gmbh, Langenhagen, D

Untersuchung der Wechselwirkungen Abfall–Wirtsgestein bei der untertägigen Ablagerung von Sonderabfällen in Salzgesteinen
U. MÜHLENWEG, Inst. f. Tieflagerung, GSF, Braunschweig, D

Das K_{oc}–Konzept zur Abschätzung der Mobilität hydrophober organischer Verbindung und zur Bewertung ihres Gefährdungspotentials
U. STARKE, Büro Dr. Hoffmann, Essen, P. GRATHWOHL, Harreß Geotechnik, Harburg, D

3 NEUENTWICKLUNGEN DER DEPONIE– UND ALTLASTENTECHNIK
3 NEW TECHNOLOGIES ON LANDFILLS AND CONTAMINATED SITES

Die Sonderabfalldeponie Rondeshagen als Beispiel für eine neue Konzeption der Abfallbeseitigung zur Vermeidung der Bildung von Sickerwasser durch eine großflächige Überdachung
H. ERNST, Ministerium f. Natur, Umwelt u. Landesentwicklung d. Landes Schleswig–Holstein, Kiel, D

Planung von Behandlungsanlagen für Sickerwässer aus Sonderabfalldeponien – Vergleich von Aufbereitungsverfahren und deren Behandlungskosten
H. LUDWIG, K.–D. WOLZ, Fichtner GmbH, Stuttgart, D

Deponieüberwachung mittels chemisch–physikalischer Industriesensoren als Zukunftsaufgabe der Geotechnik
E. REUTER, Wersche GmbH, Oldenburg, D

The Design of Remedial Action on a Built–Up Landfill
H. HIDDING, Oranjewoud BV, Heerenveen, NL

Praxis der Altlastensanierung, dargestellt am Beispiel der Sanierung einer chemischen Fabrik
H.–D. SONNEN, Harbauer GmbH, Berlin, D

Arbeiten im Bereich kontaminierter Standorte – Maßnahmen zum Schutz der Beschäftigten
M. BANDMANN, Tiefbau–Berufsgenossenschaft, München, D

Feststoffreiche Dichtwandmassen zur Sicherung von Altlasten
R. HERMANNS, ETH Zürich, CH, E. REUTER, Wersche GmbH, Oldenburg, D

How to Deposit Asbestos–Bearing Waste Correctly – Status and Prospects
J. KLEINEBERG, Karlsruhe, D

4 REINIGUNG KONTAMINIERTER BÖDEN
4 SOIL DECONTAMINATION TECHNOLOGIES

Ergebnisse der Bodenreinigung im Hochtemperaturverfahren "System–Züblin"
E. GLÄSER, Züblin AG, Stuttgart, D

Thermische Reinigung von verunreinigten Böden, eine bewährte Technik
R.C. REINTJES, R&E Consult, W.F. KOOPMANS, Ecotechniek, Utrecht, NL

Bodensanierungszentrum Bochum – Thermische Bodenreinigung
J. FORTMANN, H. KRAPOTH, Ruhrkohle Umwelttechnik GmbH, Essen, D

Die Behandlung kontaminierter Böden am Modell eines Bodensanierungszentrums
H. BEYWINKLER, SGP, Wien, A

The NBM–Indirect Heated Thermal Soil Cleaning System
H.J. van HASSELT, A. COSTERUS, NBM, The Hague, NL

Venting–Mediated Removal of Dieseloil from Subsurface Soilstrata as a Result of Stimulated Evaporation and Enhanced Biodegradation
J. van EYCK, C. VREEKEN, Delft Geotechnics, NL

Index